CONTENTS

CONTENTS (Continued)

Sensemaking
A Structure for an Intelligence Revolution

David T. Moore

NATIONAL INTELLIGENCE UNIVERSITY
WASHINGTON, DC

July 2012
Second Edition

The views expressed here are those of the author and do not reflect the official policy or position of any branch of the U.S. Government.

A. Denis Clift.

Sensemaking: A Structure for an Intelligence Revolution, 2nd Edition, David T. Moore is from our new series, The A. Denis Clift Series on the Intelligence Profession. The Clift Series presents original research on intelligence analysis and the teaching of intelligence. In 2009, A. Denis Clift concluded a 50-year career with the federal government. Clift was president of the National Defense Intelligence College (NDIC) from 1994 to 2009 and was instrumental in creating the Center for Strategic Intelligence Research, which houses the National Intelligence NI Press.

David T. Moore argues that sensemaking can be accomplished as a collaborative enterprise. Intelligence professionals can work with executive decisionmakers to explain data that are "sparse, noisy, and uncertain," but this requires an interpreter and experienced champion to bring about a practicable understanding and acceptance of the concept among intelligence practitioners.

The goal of the NI Press is to publish high-quality, valuable, and timely books on topics of concern to the Intelligence Community and the U.S. government. Books published by the NI Press undergo peer review by senior officials in the U.S. government as well as outside experts.

How to order this book. Everyone may download a free electronic copy of this book from our website at *http://www.NI-U.edu*. U.S. government employees may request a complimentary copy of this book by contacting us at: *press@NI-U. edu*. The general public may purchase a copy from the Government Printing Office (GPO) at *http://bookstore.gpo.gov*.

Editor, NI Press
Center for Strategic Intelligence Research
National Intelligence University
Defense Intelligence Agency
Joint Base Anacostia-Bolling
Washington, DC 20340-5100

ISBN	978-1-937877-00-2
GPO Sales Stock Number	008-000-01062-5
Library of Congress Control Number	2012934290

CONTENTS (Continued)

CONTENTS (Continued)

Tables

CONTENTS (Continued)

Figures

FOREWORD

Gregory F. Treverton
Director
RAND Corporation Center for Global Risk and Security

Recently, I had the opportunity to work with some very impressive young analysts at David Moore's recent home-away-from-home, the National Geospatial-Intelligence Agency (NGA). Assisting U.S. operations in Iraq and Afghanistan, they understood how much the world of what we still call "imagery" had changed. As they put it: We at NGA used to look for things and know what we were looking for. If we saw a Soviet T-72 tank, we knew we'd find a number of its brethren nearby. Now, though, we're not looking for things. Instead, we're looking for activities or transactions. And we don't know what we're looking for.

In fancier language, the paradigm of intelligence and intelligence analysis has changed, driven primarily by the shift in targets from the primacy of nation-states to trans-national groups or irregular forces. In the world of the national-state, I and others divided intelligence problems into puzzles and mysteries (or variants of those words).[1] Puzzles are those questions that have a definitive answer in principle. How many nuclear missiles the Soviet Union had was a puzzle. So is whether Al Qaeda possesses fissile material. By contrast, mysteries are questions that cannot be answered with certainty. They are future and contingent. Will North Korea reach agreement to cease its nuclear program? No one knows the answer, not even North Korean leader Kim Jong Un. It depends. The question is a mystery, not a puzzle.

For puzzles, intelligence tried to produce *the* answer. In solving puzzles about the Soviet Union, the United States spent billions of dollars, primarily on the technical systems whose fruits were and are analyzed at David's home institution, the National Security Agency (NSA) and at NGA, along with espionage collected by the CIA. For mysteries there was no answer. Instead, analysts sought to frame the mystery by providing a best estimate, along, perhaps, with excursions or scenarios to test the sensitivity of critical factors. If intelligence failed to understand the full picture of Soviet missiles,

---

[1] For my version of the distinction, see Gregory F. Treverton, "Estimating Beyond the Cold War," *Defense Intelligence Journal* 3, no. 2 (Fall 1994): 5-20. Cited hereafter as Treverton, "Estimating."

and puzzle became mystery, it at least knew something about where to look: there was experience and theory about missile building, plus historical experience of Soviet programs. The mystery came with some shape.

However, today's transnational threats confront us with something more than mysteries. I call these shapeless mysteries-plus "complexities," borrowing Dave Snowden's term. They are sometimes called, as Moore notes, "wicked problems" or simply "messes." They come without history or shape. Large numbers of relatively small actors respond to a shifting set of situational factors. Thus, they do not necessarily repeat in any established pattern and are not amenable to predictive analysis in the same way as mysteries. Those characteristics describe many transnational targets, like terrorists—small groups forming and reforming, seeking to find vulnerabilities, thus adapting constantly, and interacting in ways that may be new.

For complexities, especially, the challenge is to employ sensemaking—the term is from Michigan psychologist, Karl Weick. Exactly how to accomplish sensemaking is a task that still mostly lies before us, which makes this book such an important contribution. Sensemaking departs, as Moore notes, from the postwar tradition of Sherman Kent, in which analysis meant, in the dictionary's language, "the process of separating something into its constituent elements." Sensemaking also blurs America's bright white line between intelligence and policy, for, ideally, the two would try to make sense together, sometimes disaggregating events, sometimes aggregating multiple perspectives, always entertaining new hypotheses, all against the recognition that dramatic failure (or success) might occur at any moment.

Sensemaking is a tall order, but there is no better Sherpa for the unfamiliar terrain of this new paradigm than David Moore. He almost uniquely embodies both practice and academic scholarship. Indeed, one of the tantalizing aspects of his academic work is that, as a careful intelligence professional (and one from NSA to boot), he is very careful about classification. That means the visible trails of his practice in his scholarship are sparse, and his cases are mostly familiar ones, albeit ones often spun in new directions.

His approach to sensemaking takes us from information foraging, harvesting and marshalling into understanding. He looks at various forms of tacit knowledge, and he and the contributors report on some intriguing tests of sensemaking. Several of us who looked around the Intelligence Community in the years after September 11th noted how little use it made of formal methods or machines other than computers for sorting.[2] Worse, in some sense the Cold War practice of analysis sought to turn humans into machines by rooting out judgment, bias, hunch, stereotyping—all the things humans do best. The new paradigm makes the use of machines and method imperative, letting machines do what they do best—searching large amounts of data, remembering old patterns, and the like—while letting humans use the judgment they alone can apply. Yet the tests by Moore and his colleagues remind us that methods are critical but only if they have been tested. It turns out, for instance, that ACH, analysis of competing hypotheses, a method more frequently used now and one that has been tested, isn't all that valuable, at least not for analysts beyond the novice level.

For years I've had at hand a bumper sticker for which I lacked the car. The bumper sticker is: Intelligence cannot truly be reshaped until it reshapes its products. So long as it thinks of products primarily as words on paper (or bytes on a screen) produced by relevant experts and stovepiped by agency it is stuck in the old paradigm. Moore should not be blamed for my bumper stickers, but his emphasis on communication echoes the concern underlying it. I'd be happy if the Intelligence Community began a number of pilot projects trying to develop sensemaking, but Moore is much more ambitious: while recognizing its limitations, he'd make sensemaking the basis of intelligence. It should be.

---

2 Most striking is the work of anthropologist Rob Johnston, now on the inside: Rob Johnston, *The Culture of Analytic Tradecraft: An Ethnography of the Intelligence Community* (Washington, DC: Center for the Study of Intelligence, Central Intelligence Agency, 2005). My own version is *Assessing the Tradecraft of Intelligence Analysis* (with C. Bryan Gabbard), TR-293 (Santa Monica, CA: RAND Corporation, 2008).

COMMENTARY

Anthony Olcott, PhD
Associate, Institute for the Study of Diplomacy
Georgetown University

David Moore is right to talk of the need for an intelligence revolution. However, as Lenin learned in the 18 years that passed between publication of *The Development of Capitalism in Russia* and taking over the Winter Palace, it takes more than a diagnosis and a prescription to make a revolution. Although his is among the best, Moore's book is also but the latest addition to a groaning shelf of books devoted to intelligence and analytic reform while the companion shelf, for books on how to improve the policy process, sits dusty and all but empty. In that regard, even though Moore's discussion of the processes of analysis and how the ways we answer questions might be improved is one of the strongest in recent memory, the most valuable part of the book could well be the somewhat smaller amount of attention it devotes to the problem of how we formulate our questions in the first place.

As Moore points out, Sherman Kent and the other *ur*-fathers of intelligence took for granted that the "intelligence questions" are self-evident— foreign policy is (Walter Lippmann's words) "the shield of the republic" and strategic intelligence is (Kent's addition to Lippmann) what "gets the shield to the right place at the right time" and what "stands ready to guide the sword."[3] In that metaphor there is no room for doubt about what is threat, what defense, or indeed what receives the thrust of the sword. Even as Kent's book appeared, however, other voices were arguing that policy formation is not so self-evident or straightforward.

Moore quotes one of these voices, that of Kent's contemporary and, for a time, IC colleague, Willmoore Kendall. Kendall did not share Kent's conviction that the job of the analyst was "to stand behind [the policymakers] with the book opened at the right page, to call their attention to the stubborn fact they may be neglecting."[4] Unlike Kent, who was an unabashed elitist, Kendall was a "majoritarian," who believed that, in a democracy, all policy, foreign and domestic, could only be set by the wishes of the "50 percent plus 1" who vote

---

[3] Sherman Kent, *Strategic Intelligence for American World Policy* (Princeton NJ: Princeton University Press, 1949), p. viii. Cited hereafter as Kent, *Strategic Intelligence*.

[4] Kent, *Strategic Intelligence*, p. 182.

for a particular person, platform, or party—and the job of the analyst, therefore, was to help the "politically responsible laymen" whom that majority had elected to bring those policies into being. Although "majoritarianism" led Kendall into some positions which today seem deeply repugnant —for example, he defended racial segregation, on the grounds that this was the wish of the majority—his argument that policy is determined by beliefs, not "objective facts," is one that, had it prevailed, could probably have helped us to avoid a good number of the familiar "intelligence failures" that Moore's book enumerates.

Moore moves some way toward Kendall's position when he describes the potential that sensemaking offers as a means precisely for helping policymakers to improve how they think about policy. The collaborative processes he outlines would help analysts and policymakers alike move from the present fixation on "how things work" (the provenance of analysis) to imaginative exploration of the ways in which things *could* work (the purpose, it would seem, of policy).

It is here particularly that I would encourage Moore's readers to think about how to move this sense-making revolution closer to reality. As philosopher Denis Hilton has remarked, there is a profound difference between "causal attribution" and "causal explanation"—in his words, "attributing the 9/11 attacks to someone is not the same as explaining them to him."[5] Moore has done a deep and convincing job of diagnosing the ills of the IC, and has proposed a rich and promising cure. This, as Hilton points out, is an extended act of cognition. What lies between this book and Moore's revolution, however, is the need to have others come to the same conclusion— which, as Hilton points out, requires *communication*, not cognition.

Sixty years ago a small group of analysts—dubbed "Talmudists" for their pains—worked out a complex, sophisticated method of deriving actionable intelligence from the tightly controlled propaganda outlets of the USSR and Mao's China. This let IC sinologists spot the first signs of the Sino-Soviet split as early as April 1952, and by 1955 Khrushchev had been tagged as the likely winner in the struggle to consolidate power in the Kremlin after Stalin's death. Those early indicators, however, remained scoffed at and un-acted upon precisely because the methodology—which a colleague in the CIA compared to studying "invisible writing on slugs"[6]—was too complex and too weird to

---

5 Denis Hilton, "Causality vs. Explanation: Objective Relations vs. Subjective Interests," *Interdisciplines*, Institute of Cognitive Sciences, University of Geneva, URL: <http://www.interdisciplines.org/causality/papers/14>, accessed 1 November 2010.

6 Richard Shryock, "For An Eclectic Sovietology," *Studies in Intelligence*, vol. 8, no. 1 (Winter 1964).

be easily explained to policymakers—*who, in any case, already believed other hypotheses, and had their own "facts."*[7]

The challenge we face is the same as that which faced those Talmudists. Moore has convinced himself, and he is highly likely to convince all the analysts who read his book, that sensemaking is indeed the intelligence revolution we need. The challenge now is to communicate that to policymakers, so that they too will be willing to join David Moore's sensemaking revolution.

---

[7] Harold P. Ford, "The CIA and Double Demonology: Calling the Sino-Soviet Split," *Studies in Intelligence*, vol. 42, no. 5 (Winter 1988-1989).

COMMENTARY

Emily S. Patterson, PhD
Assistant Professor
College of Medicine
The Ohio State University

A colleague once said that he was dedicated to the vision of having decisionmaking be directly informed by evidence rather than the popularity of the latest fad or pet projects of powerful leaders. The context for his comment was for deciding what innovations to implement in hundreds of Intensive Care Units to reduce risk-adjusted patient mortality. Nevertheless, I believe that this lofty goal could easily apply to United States policymaking. In my opinion, it is an achievable goal for the vast majority of United States policy to be directly informed by evidence that is systematically validated, collated, and synthesized by teams of professional intelligence analysts.

This book is a critical milestone in attaining the goal of analysis directly supporting evidence-based policymaking. This book's primary contribution is to conduct sensemaking on the label sensemaking. Decades of relevant academic literatures have been synthesized into one framework that illustrates how disparate research streams relate to each other and to the framework. Until now, there has not been such an extensive effort to pull together related research on sensemaking from such diverse disciplines as psychology, political science, philosophy, organizational science, business, education, economics, design, human-computer interaction, naturalistic decisionmaking, and macrocognition.

The contributions of this book go beyond a literature review, however, in that an action-oriented stance is taken toward capturing nuggets of insight on how to improve aspects of analysis. The categories themselves are useful in putting some shape and structure to the amorphous value that expertise brings to creating a solid analytic product in an uncertain world: planning, foraging, marshaling, understanding, and communicating. Of particular value is describing different aspects of validation that are relevant to intelligence sensemaking, and distinguishing processes for predicting future events (foresight) from processes for describing past events and assessing their impacts (hindsight).

Another colleague once said that she looks at what is measured operationally to determine how people truly define a concept. Chapter 8, Establishing Metrics of Rigor, is therefore critically important to any discussion of how to encourage improved sensemaking in the Intelligence Community. As David Moore notes, although we believe our rigor metric to be a promising first step, much more needs to be done to ensure that all of the important aspects of rigorous analysis are captured. The application of the rigor metric to a face valid case study in this book is the first such application in intelligence analysis to compare processes by different analytic teams. Whether or not high rigor on all of these inter-related dimensions is possible to achieve under the working conditions for intelligence analysts today is an unresolved debate. Even if high rigor is not possible under extreme time pressure, data overload, and workload conditions, the measure has potential value in supporting negotiations for what aspects are most important to do well for a given task, as well as communicating the strengths and weaknesses of the process behind an analytic conclusion.

COMMENTARY

Christian P. Westermann
Senior Analyst
Bureau of Intelligence and Research
U.S. Department of State

History will tell us if current intelligence reforms are evolutionary or revolutionary, but the Intelligence Community is responding to mandated change brought about by the 2004 Intelligence Reform and Terrorism Prevention Act (IRTPA).[8] In particular, the analytic and collector communities are adjusting to one of IRTPA's pillars—improved information sharing. As reforms unfold, the collector and analyst must adapt to new rules and new analytic standards, and incorporate more methodologies, techniques, and alternatives in their analysis, in collaboration with managers and tradecraft cells in the national intelligence organizations. These new structures and guidelines present an intellectual challenge as well as a bureaucratic maze for the collector and analyst struggling not only to "produce" intelligence in a timely fashion but also to improve their product. This is not easy for the intelligence professional because time is not on their side. This is why improving the way in which all analysts think is so important and why an understanding of sensemaking will help advance the profession beyond the "established analytic paradigm" for complex problems and create greater possibilities for the application of imagination in the IC. The failure to properly assess Saddam Hussein's WMD programs during the lead-up to Operation Iraqi Freedom is the preferred example of this failure to imagine alternatives. The corporate solution to this problem is increased collaboration and information sharing; David Moore is not in disagreement but has suggested that it must go beyond new methodologies or techniques—it must be done with a strong sense of rigor and individualism in one's thinking.

David Moore has written for the Intelligence Community a revolutionary epistemology His novel construct for intelligence professionals is the foundation for a philosophy of intelligence. He has started where he left off in his work on *Critical Thinking and Intelligence Analysis* with a path forward for analysts and how they can improve their capacity and move beyond the

---

8 United States Congress, *Intelligence Reform and Terrorism Prevention Act of 2004*, 108th Congress, 2nd Session, 20 January 2004. Cited hereafter as U.S. Congress, IRTPA, 2004.

methods they learn as "good" tradecraft.[9] Moore's prescription is to take the disaggregation of data, commonly referred to as analysis, synthesize it, and then apply to it one's interpretation and communication skills to make sense of the information. Sensemaking therefore is a theory of knowledge for the intelligence professional and also a practice to aid the difficult art of intelligence reasoning.

Current "total" intelligence reform, as described by Director of National Intelligence James Clapper in late 2010, is reform that is focused on "integration, the merging of collection and analysis—particularly at the ODNI level—analytic transformation, analytic integrity, acquisition reform, counterintelligence—and information sharing."[10] This involves a great deal of uncertainty for organizations and analysts, in light of the formation of new intelligence fusion centers reminiscent of Defense intelligence reform of the 1990s—which saw the rise of Joint Intelligence Centers—and presents a challenge to make sense of new security challenges in the post-9/11 world. The time is therefore ripe for analysts to transform their thinking and tradecraft and Moore's new paradigm offers real improvements to their practice of intelligence. His attention to revolutionary change in the art of intelligence thinking grows from his recognition that organizational reform has been ongoing for decades and despite those changes attendant failures have occurred and continue to occur. Therefore the only hope for achieving positive reform rests with changing the practice of intelligence whereby the individual collector and analyst, working together, and accepting the responsibility to think critically but also independently and across the Community, make sense of the 21st century national security environment.

9 David T. Moore, *Critical Thinking and Intelligence Analysis* (Washington, DC: NDIC Press, 2007). Cited hereafter as Moore, *Critical Thinking*.

10 James Clapper, "Remarks and Q & A by Director of National Intelligence Mr. James Clapper," Bipartisan Policy Center (BPC)—The State of Domestic Intelligence Reform, 6 October 2010, URL: <http://www.dni.gov/speeches/20101006_speech_clapper.pdf>, accessed 29 October 2010.

COMMENTARY

Phil Williams, PhD
Director, Matthew B. Ridgway Center for International Security Studies
Wesley W. Posvar Chair of International Security
Graduate School of Public and International Affairs
University of Pittsburgh

Moore's Law for Intelligence

Any book that discusses amongst other things, red brains and blue brains, kayaking, information foraging, flashlights as blindfolds, space-time envelopes, and intellectual audit trails, is out of the ordinary. When you throw in the contention by the author that intelligence as currently practiced is akin to medicine in the 14th Century you have a book that will raise hackles, blood pressure, and voices. David Moore's provocative and stimulating analysis of critical thinking and sensemaking for intelligence does all of the above.

This is not an easy read. But the overall thesis is straightforward and compelling: the environment within which the U.S. Intelligence Community now finds itself is not only highly complex but also full of wicked problems. To provide the kind of intelligence that is useful, relevant, and helpful to policy makers who have to anticipate and respond to these problems and challenges, Moore argues that the traditional paradigm developed largely by Sherman Kent has to be superseded by a new paradigm based largely on ideas initially outlined by Willmoore Kendall, a contemporary critic of Kent. The original Moore's Law[11] was narrowly technical; David Moore in contrast argues that a complex environment full of mysteries, not puzzles, requires holistic thinking (as opposed to simply disaggregation of problems), mindfulness (as opposed to mindlessness which he also elucidates), and a dynamic willingness to change paradigms, shift perspectives, and abandon strongly held perceptions. The book also develops the notion of sensemaking rigor and shows how metrics of rigor can be applied to several studies examining the rise and impact of non-state actors.

David Moore's analysis is important and deserves to be widely read in the Intelligence 0 and in the academic world. Yet, the volume—as he would

---

11 Gordon E. Moore, "Cramming More Components onto Integrated Circuits," *Electronics*, vol. 38, no. 8 (19 April 1965), URL: <ftp://download.intel.com/museum/Moores_Law/Articles-Press_Releases/Gordon_Moore_1965_Article.pdf>, accessed 7 November 2010.

be the first to acknowledge—is intended as an early word on sensemaking and intelligence, rather than the last word. Indeed, it would have been helpful, for example, if David Moore had considered more explicitly the argument by David Snowden that making sense of a complex environment requires probing the environment. Further thought about this suggests that law enforcement is particularly good at this form of knowledge elicitation and sensemaking: sting operations, controlled deliveries, infiltration of criminal organizations, are all probing mechanisms that can contribute significantly to an increased level of understanding and, concomitantly, to an enhanced capacity for effective action. For many intelligence professionals, especially those who have had a dismissive view of law enforcement, the idea that law enforcement approaches to sensemaking might be ahead of those in the Intelligence Community, is likely to be as uncomfortable as most of the arguments in David Moore's book. Certainly Moore's volume is designed to shake and to stir. It is a manifesto for an intellectual revolution in the approach to intelligence and, as such, is likely to be both acclaimed and reviled. One suspects that the author will measure his own success by the depth of opposition as well as the levels of support for the revolution he is proposing.

PREFACE
On Being Mindful

What Is Mindlessness?

We are surrounded by errors and they are ours. Intelligence officials at the national level repeatedly use the same excuses for professional errors and for the systemic failures that follow. Despite directives to "fix" the structures, and most recently the means, by which intelligence is created, we insistently fail at our obligation to make early sense of vital threats and opportunities. What is our problem?

We may begin by examining the concept of *mindlessness*. Ellen Langer, summarizing her pioneering social psychology research, finds mindlessness to arise from an over-reliance on "categories and distinctions created in the past."[12] She holds that such categories "take on a life of their own."[13] Seen in this context, the failure to recognize in foresight that an American or a Nigerian man could be a member of Al Qaeda arises from a rigid deductive categorization of who is a member of Al Qaeda and is a case of mindlessness.

Langer also sees mindlessness arising from "automatic behavior." Here, people rely on automatic responses as the basis for their behavior, as when one writes "a check in January with the previous year's date."[14] By extension, intelligence professionals, in assessing sources, may develop a habit of discounting human intelligence sources because some are untrustworthy. As a result, they may miss novel insights because they use certain sources to the exclusion of others.

Finally, mindlessness can result from a failure to take into account alternative information that transcends our comfortable worldview. Langer observes that "[highly] specific instructions…encourage mindlessness" because they define what is acceptable and limit the viability of alternative signals that could lead to more accurate understanding of a phenomenon.[15] During the summer of 1962, CIA and DIA intelligence professionals at the Refugee

---

[12] Ellen J. Langer, *Mindfulness* (Cambridge, MA: Da Capo Press, 1989), 11. Cited hereafter as Langer, *Mindfulness*.
[13] Langer, *Mindfulness*, 11.
[14] Langer, *Mindfulness*, 12.
[15] Langer, *Mindfulness*, 17-18.

Processing Center (RPC) in Miami discounted the validity of refugee reports of nuclear missiles on the island because some refugees also went beyond the pale by concocting "farfetched tales of African troops with rings in their noses, lurking Mongolians, and even Chinese troops."[16] There were in fact 100 nuclear-tipped *tactical* missiles deployed on the island months before the arrival of the more infamous strategic missiles.[17] A rigid notion of what constituted a nuclear missile, usually conceived as an offensive weapon, appears to have contributed to the case officers' mindless disregard of the witnesses.

Not only intelligence practitioners, but also those with whom they communicate their understandings, remain subject to the dangers of mindlessness. With respect to intelligence consumers, two faculty members at the International Institute of Management Development (IMD), corporate strategy expert Cyril Bouquet and corporate leadership and organization expert Ben Bryant, suggest that "decision makers often suffer from poor attention management, being obsessed with the wrong types of signals and ignoring possibilities that could significantly improve the fate of their undertakings."[18] They characterize these behaviors as *fixation* and *relaxation*. People who fixate "become so preoccupied with a few central signals that they largely ignore things at the periphery."[19] A deadly example of fixation occurred in 1977, when the KLM 747 flight crew on Tenerife failed to avoid a collision with a Pan Am 747: "they didn't give sufficient attention to the presumably very important communications coming in from air traffic controllers."[20] Naval aviators engaged in night landings on aircraft carriers also fixate and this leads to pilots' ignoring "the obvious and doing the inexplicable" (and

---

[16] James H. Hansen, "Soviet Deception in the Cuban Missile Crisis," *Studies in Intelligence,* 46, no. 1 (2002), 56. The author referred to this incident in his book on critical thinking and intelligence. See Moore, *Critical Thinking,* 20.

[17] Raymond L. Garthoff, "US Intelligence in the Cuban Missile Crisis," in James G. Blight and David A. Welch, eds., *Intelligence and the Cuban Missile Crisis* (London, UK: Frank Cass, 1998), 22.

[18] Cyril Bouquet and Ben Bryant, "The Crisis Is Here To Stay. Do You Have The Key To Coping?" *Forbes,* 21 April 2009, Online Edition, URL: <http://www.forbes.com/2009/04/21/stress-coping-mindfulness-leadership-managing-fixation.html>, accessed 13 January 2010. Cited hereafter as Bouquet and Bryant, "Key to Coping." IMD is not well known within the Intelligence Community. Its Executive Education program has been rated the number two program worldwide by the *Financial Times* for the past three years; *The Economist* rates its MBA Program as the second best in the world. See *Financial Times,* "Executive Education—open—2009, URL: <http://rankings.ft.com/businessschoolrankings/executive-education—open>, accessed 15 January 2010; and *The Economist,* "Which MBA? 2009 Full-time MBA Ranking," URL: <http://www.economist.com/business-education/whichmba/>, accessed 15 January 2010.

[19] Bouquet and Bryant, "Key to Coping."

[20] Bouquet and Bryant, "Key to Coping." The accident occurred when KLM 747 collided with the Pan Am 747 as the former took off over the latter. In the crash 583 passengers and crew were killed although miraculously, 65 passengers and crew on the Pan Am flight survived.

often dying).[21] Such fixations are goal-oriented—taking off in the former example and landing in the second. The pilots know intellectually what to do but their fixation on an emotional goal mis-focuses them: "knowing [is] no match for emotion."[22] While they are fixated, they are not truly aware. Nor are they living in the moment; instead, they are inadvisably envisioning themselves already in the air or on the deck and expecting the resulting emotional sense of relief.

Bouquet and Bryant identify relaxation as when, after a "sustained period of high concentration," people become unfocused on the task at hand and look to the ultimate goal.[23] A case in point involves the deaths of three climbers on Oregon's Mount Hood in 2002: they were distracted from the matter at hand—that of completing a difficult descent; they took shortcuts.[24]

Both fixation and relaxation contribute to intelligence failures. For intelligence practitioners, focusing on the wrong factors and failing to recognize the significance of novel indicators are examples of fixation that may have been at work in the December 2009 failure to anticipate the attempted bombing of Northwest Airlines flight 253 over Detroit. Sometimes ascribed to intelligence professionals' and national consumers' falling prey to "creeping normalcy," relaxation was also a contributor to Israel's failure to anticipate the attacks by Egypt and Syria in 1973.

In sum, mindlessness too often guides the assessment of affairs in too many domains, leading to errors, failures, and catastrophes. Mindlessness is deemed unacceptable within the larger American society only when the resulting errors do lead to accidents and disasters. However, mindlessness is completely unacceptable within the domain of intelligence. One can never be certain in foresight whether errors will occur, so intelligence professionals must seek to anticipate, recognize and avoid them at all costs.

Attaining Mindfulness

The antithesis of mindlessness is mindfulness.[25] For Langer, a mindful state corresponds with: "(1) [aptitude for the] creation of new categories;

---

21 Laurence Gonzales, *Deep Survival: Who Lives, Who Dies, and Why* (New York, NY: W.W. Norton, 2003), 26. Cited hereafter as Gonzales, *Deep Survival*.

22 Gonzales, *Deep Survival*, 35.

23 Bouquet and Bryant, "Key to Coping."

24 Gonzales, *Deep Survival*, 97-123.

25 One of the origins of the concept of mindfulness lies in the work of the 19th-century U.S. philosopher Charles Sanders Peirce, whose notion of "thirdness" relates to present connotations of "mindfulness" by recognizing thirdness as an observer's self-referenced content of an interpretation. See: Charles Sanders Peirce, "Lowell Lectures (1903)," *Collected Papers of Charles Sanders Peirce*, Volume 1, Charles Hartshorne and Paul Weiss, eds. (Cambridge, MA: Harvard University Press, 1958), para 331-332.

(2) openness to new information; and (3) awareness of more than one perspective."[26] For example, as an intelligence professional considers who might be a member of Al Qaeda, a *mindful* attitude would involve constant reassessment and categorization of who might hold such membership—leaving the path open to new information for making sense of the organization and its membership. Thus, as we apply the idea that "[a] steer is a steak to a rancher, a sacred object to a Hindu, and a collection of genes and proteins to a molecular biologist,"[27] the notion of a Nigerian male or even a blonde woman from Pennsylvania as a possible Al Qaeda affiliate would emerge from a mindful perspective.

Leadership scholar Deepak Sethi sees mindfulness as a "form of meditation" teaching "three simple-on-the-surface yet revolutionary skills: Focus, Awareness, and Living in the Moment."[28] This definition descends from millennia of Buddhist tradition. He argues that rather than an esoteric method it is "very practical, action oriented, and transformational." Sethi believes that one practical way to bring about mindfulness is through the use of daily meditation, first using one's breathing as a focus, and then using "specific daily activities such as meetings with another colleague." However, the "real challenge [of employing mindfulness] is to take it from the meditation chair to the office chair and the real world."[29] Intelligence journeymen face this dilemma from a different perspective. They confront the real world and are challenged to contemplate their own thought processes as they engage it.

Australian expert on argumentation Tim Van Gelder distinguishes mindfulness from critical thinking (which he calls "metacognition"): "Metacognition is concerned with what you are thinking about. Mindfulness is concerned with *how* you think as you go about what you are doing."[30] While Van Gelder accepts Ellen Langer's definition, he argues that metacognition and mindfulness can bear an inverse relationship to one another. As an individual masters one, the need for the other diminishes. To illustrate this point, he asserts that "[in metacognitive terms, a novice] driver needs to pay lots of attention to even the most mundane aspects of driving, such as where the gearshift is. The experienced [mindful] driver pays very little attention to

---

26 Langer, *Mindfulness*, 62.

27 Langer, *Mindfulness*, 69.

28 Deepak Sethi, "Mindful Leadership," *Leader to Leader*, no. 51 (Winter 2009), 7. Cited hereafter as Sethi, "Mindful Leadership."

29 Sethi, "Mindful Leadership," 10.

30 Tim Van Gelder, "Mindfulness Versus Metacognition, and Critical Thinking," [Weblog Entry] *Bringing Visual Clarity to Complex Issues*, 27 May 2009, URL: <http://timvangelder.com/2009/05/27/mindfulness-versus-metacognition-and-critical-thinking/>, accessed 13 January 2010. Emphasis in original. Cited hereafter as Van Gelder, "Mindfulness."

driving, and can carry on a lively conversation instead."[31] Van Gelder's viewpoint—which derives from his definition of critical thinking—holds a serious implication for the intelligence professional: if one learns to be mindful, and is good at it, the risk of a lapse in critical thinking will have increased.

For example, empirical studies of the level of distraction among drivers who are using cell phones reveal an associated, *diminished* driver capacity.[32] Non-distracted drivers are receiving information—without conscious awareness—from: traffic changes ahead, behind, and along side; various mirrors in and on the car; and weather and light conditions. All these inputs factor into their decisions on how to proceed. They are intuitively "in the moment," that is, *mindful* of what they are doing and what is going on around them. They are, as Sethi notes, focused, aware, and living in the moment. But if, as experienced drivers, they are at ease in carrying on a telephone conversation, are they not more accident-prone as a result of operating with a reduced capability or likelihood to be thinking critically?

Van Gelder notes that novice drivers likely monitor their thinking and actions closely—at least intermittently.[33] The amount of metacognitive monitoring that occurs as they become more skilled probably diminishes under normal circumstances. However, it is the author's experience, drawn from office and classroom observation, that people who are skilled critical thinkers still tend to be able to question what is occurring around them even as they are aware of how they are thinking about it. They are thinking critically, even if it is not obvious that they are doing so.

Drivers who employ critical thinking skills to ensure they remain mindful of the appropriate stimuli continually make sense of the environment in which they find themselves. Being metacognitively aware of their likely diminished capacity to drive safely while engaging in cell phone conversations (as well as texting), they are likely to engage in these dangerous acts less often than one who is mindlessly fixated on the cell phone conversation and not mindfully aware of the environment inside and outside their vehicles. To use Gelder's definitions, they are thinking about what to do as well as how to do it; critical thinking and mindful thinking inform each other.

Ben Bryant and IMD research associate Jeanny Wildi write that mindfulness "involves the ability to accurately recognize where one is in one's

---

31 Van Gelder, "Mindfulness."

32 For more information on diminished driver capacity see Transportation Research Board of the National Academies, "Selected References on Distracted Driving: 2005-2009," URL: <http://pubsindex.trb.org/DOCs/Publications from TRIS on Distracted Driving.pdf>, accessed 9 December 2009.

33 Van Gelder, "Mindfulness."

emotional landscape and allows…understanding, empathy, and capacity for accurate analysis and problem-solving."[34] They identify a process of detaching, noticing, and developing "here and now awareness."[35] Detachment, for example, allows a viewer to remember that a movie is really merely a "beam of light passing through a piece of moving celluloid projecting onto a screen with some sound and music that are designed to generate particular emotions."[36]

In intelligence work, detachment involves stepping back from the full sensual experience of an issue to consider the actors involved, their motives, the larger context. Critical thinking as it is taught in the Intelligence Community attempts to make sense of the overall purpose or goal of a phenomenon, the points of view and assumptions of the actors involved, the implications of their acting in certain fashions, and other aspects of the larger context surrounding the issue.[37] Questioning the available evidence and the inferences arising from it brings further detachment from the issue.

Noticing involves remaining open to both internal and external stimuli. Ultimately, situational information is conveyed from external sources through sight, sound, touch, smell, and taste. People can think consciously about these but they tend to process them using more autonomic brain structures, often without noticing they are doing so. The unease one feels about getting into a taxi or onto an elevator in an unfamiliar setting are examples of such input. In intelligence work this might be represented as a hunch about what an adversary will do. As Daniel Kahneman and Gary Klein note, in certain environments—where one can learn the cues—these intuitions may be quite accurate.[38] However, in domains where one has not developed expertise, such intuitions can be *inaccurate*.[39] The challenge is determining which of these situations one is in. This brings us back to the imperative of applying mindful detachment from the situation.

---

[34] Ben Bryant and Jeanny Wildi, "Mindfulness," *Perspectives for Managers*, no. 162 (September 2008), 1, URL: <http://www.imd.ch/research/publications/upload/PFM162_LR_Bryant_Wildi.pdf>, accessed 14 January 2010. Cited hereafter as Bryant and Wildi, "Mindfulness."

[35] Bryant and Wildi, "Mindfulness," 2-3.

[36] Bryant and Wildi, "Mindfulness," 3.

[37] These are some of the "Elements of Thought" developed by Linda Elder, Richard Paul, and Gerald Nosich of the Foundation for Critical Thinking. This author operationalized the Foundation's critical thinking paradigm for intelligence work. See Moore, *Critical Thinking*, 8-9; and Gerald M. Nosich, *Learning to Think Things Through: A Guide to Critical Thinking Across the Curriculum*, 3rd edition (Upper Saddle River, NJ: Pearson-Prentice Hall, 2009), 50-67. Cited hereafter as Nosich, *Learning to Think Things Through*.

[38] Daniel Kahneman and Gary Klein, "Conditions for Intuitive Expertise: A Failure to Disagree," *American Psychologist*, vol. 64, no. 6 (September 2009), 520. Cited hereafter as Kahneman and Klein, "Intuitive Expertise."

[39] Kahneman and Klein, "Intuitive Expertise," 521-522.

Critical thinking assists noticing as well. Bryant and Wildi observe that because people tend to make immediate sense of stimuli, "We all too often contaminate our perceptions with unexamined assumptions we have already internalized."[40] In other words, from the time when we have already decided what something means, everything we notice tends to confirm that interpretation. However, by challenging critically our assumptions and choices of evidence (and their interpretations) and asking about alternative explanations, we may bring our noticing back to a state where it "contributes to mindfulness by keeping us open to our experiences of the external."[41]

Bryant and Wildi consider "here and now awareness" to involve paying attention to immediate experience—as it happens. Clausewitz observes this phenomenon when he distinguishes between a commander's plan for war and the friction of war.[42] The former must be adapted in light of the latter. Knowing how to do this, and when, requires the situational awareness found in the here and now. As has been noted, critical thinking allows questioning of both what one is doing and how one is doing it. Here again, one can challenge whether one is focused in the present or dwelling in the past or imagining a future. One can ask whether one is relaxed or fixated on a goal; we can also reflect on whether we are acting mindlessly or mindfully. The key is remaining vigilant, as Warren Fishbein and Gregory Treverton note, as "[mindfulness] is the result of a never-ending effort to challenge expectations and to consider alternative possibilities."[43]

Such mindful vigilance can be a lifesaver, as *Foreign Policy* columnist and blogger Thomas Ricks notes about its absence among troops serving in Afghanistan, where Marines wearing iPods while on patrol can (and apparently do) fail to notice changes in the environment they have previously patrolled, and get hit by improvised explosive devices.[44] If they were mindful, they could have noticed that a hole in the road when they went out on patrol has been filled in before their return. Instead, the "turret gunner [is]

---

[40] Bryant and Wildi, "Mindfulness," 3.

[41] Bryant and Wildi, "Mindfulness," 3.

[42] Carl von Clausewitz, *On War*, COL James J. Graham, trans. (London, UK: Keegan Paul, Trench, Truebner & Co., Ltd.: 1908), 80.

[43] Warren Fishbein and Gregory Treverton, "Making Sense of Transnational Threats," *Sherman Kent Center Occasional Papers*, vol. 3, no. 1 (October 2004), 17, URL: <https://www.cia.gov/library/kent-center-occasional-papers/pdf/OPV3No1.pdf>, accessed 8 November 2010. Cited hereafter as Fishbein and Treverton, "Making Sense."

[44] Thomas E. Ricks, "A Marine's Afghan AAR (XIV): Get Rid of the iPods on Patrol," Web Log, Foreign Policy. URL: <http://ricks.foreignpolicy.com/posts/2010/01/20/a_marine_s_afghan_aar_xiv_get_rid_of_ the_ipods_on_patrol>, accessed 26 January 2010. Ricks quotes a Marine Corps source, referred to in the web log as "CWO2/Gunner Keith Marine." Cited hereafter as Ricks, "Get Rid of the iPods." iPod is a registered trademark of Apple Computer.

watching a movie on an iPod…[and] the back seater [is] listening to music on his [iPod]."[45] Not only are they not mindful, but no one is paying attention. According to Rick's Marine informant, ensuing and repeated iPod-wearing-induced mindlessness led to the unit's losing "15 out of 20 vehicles in about a month."[46] Clearly, the cost of not being mindful is high. However, becoming mindful is not difficult; that cost is relatively low.

Developing mindfulness is not—as Bouquet and Bryant, as well as Sethi, observe—an arcane spiritual practice. The process certainly involves self-reflection—meditation to some—but in fact self-reflection has become a central facet of professional practice in "real-world" security and defense planning.[47] However, mindfulness in operations or in intelligence is neither a panacea nor a formula: Bouquet and Wildi observe that "executives need to meditate in their own way, find ways to step back and reflect on their thoughts, actions, and motivations, and decide which ones are really supportive of their strategic agendas."[48] One of the benefits of such meditation, according to recent experimental findings, is that perceptual sensitivity and vigilance improve in situations requiring sustained visual attention.[49] While percep-tual sensitivity and increased vigilance could also be attained, as in the case cited by Ricks, by simply leaving the iPods turned off and actually conducting reconnaissance and surveillance of one's surroundings, the problem of sus-tained-attention failure remains. Simply put, mindfulness declines over time. However, as MacLean *et alia* have demonstrated, sustained-attention failure can be reduced through formal meditation training.[50] In this case, formal

---

[45] Ricks, "Get Rid of the iPods."

[46] Ricks, "Get Rid of the iPods."

[47] For an example of a "how-to" guide to reflective practice, see Faculty of the School for Advanced Military Studies, *The Art of Design*, vol. 2 (Fort Leavenworth, KS, 2010). Available at URL: <http://www.cgsc.edu/events/sams/ArtofDesign_v2.pdf>, accessed 20 May 2010. This student text, which aims to prepare senior military officials for leadership roles in overseas operations, employs "reflective practice [to construct] a cognitive framework for how to rea-son through complexity." The doctrinal publication develops and applies reflective thinking ideas originated by Donald Schön. See URL: <http://www.infed.org/thinkers/et-schon.htm>, accessed 20 May 2010.

[48] Bouquet and Bryant, "Key to Coping."

[49] Katherine A. MacLean and others, "Intensive Meditation Training Improves Perceptual Discrimination and Sustained Attention," *Psychological Science*, vol. 21, no. 6 (2010), 829. Cited hereafter as MacLean et al., "Intensive Meditation." See also, John Cloud, "Losing Focus? Studies Say Meditation May Help," *Time*, online edition, 6 August 2010, URL: <http://www.time.com/time/health/article/0,8599,2008914,00.html>, accessed, 11 August 2010. Cited hereafter as Cloud, "Meditation."

[50] MacLean et al., "Intensive Meditation," 829.

meditation practices appear to sustain longer-term mindfulness—something the U.S. Army hopes will enhance the capabilities of its soldiers.[51]

While the U.S. Military and the Intelligence Community (like business enterprises) can provide environments conducive to developing mindfulness, Bouquet and Bryant remind us that developing mindfulness is "the responsibility of individuals, not companies."[52] The simple expedient of not engaging in mindfulness-reducing activities is one means of enhancing mindfulness; there are many others.[53] Critical thinking provides one self-reflective or *metacognitive* means to ascertain what surrounding phenomena are or are not taken into account.[54] Such mindfulness in turn supports the larger objective of intelligence sensemaking, the subject of this book. The author is aware of the confounding problem that intelligence may need to give attention to the entire domain of human behavior because every sphere of human practice and knowledge can be of interest and of use in the process of sensemaking. To limit this challenge, this book focuses on those areas that in the author's observation and experience appear most germane to the successful practice of national intelligence. The author intends for this book to generate beneficial discussion and further consideration of exactly what it means to engage in intelligence sensemaking and how one can go about it effectively.

---

[51] Bonnie Rochman, "Samurai Mind Training for Modern American Warriors," *Time*, online edition, 6 September 2009, URL: <http://www.time.com/time/nation/article/0,8599, 1920753,00.html>, accessed 4 August 2010.

[52] Bouquet and Bryant, "Key to Coping."

[53] Cloud, "Meditation."

[54] "Metacognitive" as used here refers to a process of critically monitoring one's reasoning about an issue while one is engaged in that reasoning or critical thinking. This makes explicit the process of reasoning. However, the term "metacognitive" refers to much more as will be developed further in the paper that follows.

ACKNOWLEDGMENTS

The author's first notes about what 21st Century intelligence would look like as well as how it might be accomplished were crafted at a 2001 conference on the future of intelligence in Italy. The intervening years brought work on what it takes to be a successful intelligence professional and how critical thinking could aid the consideration of strategic issues and the creation of intelligence knowledge for decisionmakers. These steps, it turned out, led to a larger vision of what the practice of intelligence could be: a process of sensemaking.

Few projects are created without substantial assistance from others. This book is no exception. Elizabeth Moore offered valuable ideas and challenged my thinking. She also once again put up with a distracted husband who worked (very) late into the night and on the weekends. Colleagues and peers read and commented on various drafts, criticizing and offering valuable suggestions that sent the author in new directions. Among them are James Bruce, Science Applications International Corporation (SAIC); Stu Card, Palo Alto Research Center; Steven Carey, NDIC; Bruce Chew, Monitor Group; Jeffrey Cooper, SAIC; Linda Elder, Enoch Hale, Gerald Nosich, and Richard Paul, Foundation for Critical Thinking; Norman Endlich, University of Maryland; Warren Fishbein, Department of State; Robert Heibel, Mercyhurst College Institute for Intelligence Studies (MCIIS); Noel Hendrickson, James Madison University; Richards J. Heuer, Jr., Central Intelligence Agency (retired); Robert Horn, Stanford University; Francis J. Hughes, NDIC; Rob Johnston, Center for the Study of Intelligence; Gary Klein, Klein Associates; Lisa Krizan, National Security Agency (NSA); Martin Krizan, NSA; Mark Lowenthal, president of the Intelligence and Security Academy, LLC; Brandon S. Minnery, IARPA (Intelligence Advanced Research Projects Activity); Esther Neckere, National Geospatial-Intelligence Agency; William Nolte, University of Maryland; Anthony Olcott, Georgetown University; Emily S. Patterson, Ohio State University; Randy Pherson, Pherson Associates; Peter Pirolli, Palo Alto Research Center; Timothy Smith, Office of Naval Intelligence; David Snowden, Cognitive Edge; Mark Stefik, Palo Alto Research Center; Kenneth Stringer, Booze Allen Hamilton; Cathryn Thurston, NI Press; Gregory Treverton, RAND; Kristan J. Wheaton, MCIIS; Kendall White, Washington and Lee University; Phil Williams, University of Pittsburgh; Chris Westermann, Department of State; David Woods, Ohio State University; and Daniel Zelik, Ohio State University.

The participants in a number of workshops sponsored by IARPA, the Office of the Director of National Intelligence (ODNI), NSA, and the ODNI Global Futures Forum also provided input and ideas. The staff of the Foreign Denial and Deception Committee's denial and deception program at the National Defense Intelligence College (NDIC) invited the author to defend an early version of the text as part of an advanced studies program.

Contributors Robert Hoffman (Chapter 5) and Elizabeth Moore, William Reynolds, and Marta Weber (Chapter 7) graciously agreed to share their ideas in collaborative efforts to bring those chapters about. Without their assistance those two chapters would have been far less rich. Colleagues of Dr. Reynolds as well as Kristan Wheaton and his students in the MCIIS (acknowledged separately in Chapter 7) provided much of the "heavy lifting" that made the conclusions derived possible.

Russell Swenson, under contract to the NDIC's Center for Strategic Intelligence Research, once again provided the essential input of a good editor. His reviews tightened arguments and focused what follows. He also provided a wealth of new ideas that sent the author off in new directions and enriched the book significantly. William Spracher once more provided the skills of a technical editor. Cathryn Thurston, the Director of the NDIC Center for Strategic Intelligence Research and NDIC Press editor, supported the creation of this project as part of her advocacy that intelligence needs to be recast for this century.

While the author was privileged to receive the assistance of others, this remains his work and any errors are his. The author may be reached via electronic mail at *david.t.moore@ugov.gov*.

Definitions for Making Sense of Sensemaking

Intelligence is a "specialized form of knowledge...[that] informs leaders, uniquely aiding their judgment and decision-making." It is a type of knowledge created through organized activity that adds unique value to the policy- or decisionmaker's deliberations. In the U.S. context, it makes sense of phenomena of interest to national leaders, warfighters, and those that directly and indirectly support them. Intelligence makes sense of phenomena related to the social behaviors of others. It reflects interest in what anyone will do to, and with, others that could affect the national interests of the United States as well as the prosperity and security of its citizens. Intelligence maintains an interest in external phenomena, such as epidemic or pandemic diseases, that impact U.S. national interests. In contrast to some popular portrayals, it really is not voyeuristic: what others do privately and alone is generally of little interest or value except as it affects how they relate to, and behave toward others. In other words, when private behaviors reveal either vulnerabilities or preferences, they may become of value to intelligence practitioners.

Sources: David T. Moore, *Critical Thinking and Intelligence Analysis, Occasional Paper Number Fourteen* (Washington, DC: National Defense Intelligence College, 2006), 2. Cited hereafter as Moore, *Critical Thinking*. Intelligence also refers to activity and organization: See Kent, *Strategic Intelligence*.

Sensemaking as it is used here refers to "a set of philosophical assumptions, substantive propositions, methodological framings, and methods." As Mark Stefik notes (referring to work done with colleagues Stuart Card and Peter Pirolli), it "is how we gain a necessary understanding of relevant parts of our world. Everyone does it." Sensemaking goes beyond analysis, a disaggregative process, and also beyond synthesis, which meaningfully integrates factors relevant to an issue. It includes an interpretation of the results of that analysis and synthesis. It is sometimes referred to as an approach to creating situational awareness "in situations of uncertainty." Gary Klein, Brian Moon, and Robert Hoffman consider the elements of sensemaking and conclude that it "is a motivated, continuous effort to understand connections (which can be among people, places, and events) in order to anticipate their trajectories and act effectively."

These authors conclude that "the phenomena of sensemaking remain ripe for further empirical investigation and [warn] that the common view of sensemaking might suffer from the tendency toward reductive explanation." By reductive explanation Klein, Moon, and Hoffman refer to a tendency to overly simplify explanations—to "reduce" complex phenomena to simplistic models facilitating an apparently needed but shallow understanding.

Sources: Brenda Dervin, "Sense-Making Methodology Site", URL: <http://communication.sbs.ohio-state.edu/sense-making/>, accessed 12 September 2007; Mark Stefik, "The New Sensemakers: The Next Thing Beyond Search Is Sensemaking," *Innovation Pipeline* (15 October 2004), URL: <http://www.parc.com/research/publications/files/5367. pdf>, accessed 11 March 2009; Dennis K. Leedom, *Final Report: Sensemaking Symposium*, 23-25 October 2001, Command and Control Research Program Office of the Assistant Secretary of Defense for Command, Control, Communications and Intelligence. Gary Klein, Brian Moon, and Robert R. Hoffman, "Making Sense of Sensemaking 1: Alternative Perspectives," *IEEE Intelligent Systems* vol. 21, no. 4 (July/August 2006), 71, 72. Cited hereafter as Klein, Moon, and Hoffman, "Making Sense of Sensemaking 1." Paul J. Feltovich, Robert R. Hoffman, Axel Roesler, and David Woods, "Keeping It Too Simple: How the Reductive Tendency Affects Cognitive Engineering," *IEEE Intelligent Systems* vol. 19, no. 3 (May/June 2004).

Intelligence sensemaking encompasses the processes by which specialized knowledge about ambiguous, complex, and uncertain issues is created. This knowledge is generated by professionals who in this context become known as Intelligence Sensemakers.

These terms are used as defined here throughout this book.

Sensemaking: A Structure for an Intelligence Revolution[55]

David T. Moore

Knowledge welcomes challenges.

— Peter Kosso

Venture boldly into nonsense. Nonsense is nonsense only when we have not yet found that point of view from which it makes sense.

— Gary Zukav

CHAPTER 1
Introduction

Where We Are

How people notice and make sense of phenomena are core issues in assessing intelligence successes and failures. Members of the U.S. Intelligence Community (IC) became adept at responding to certain sets of phenomena and "analyzing" their significance (not always correctly) during the Cold War. The paradigm was one of "hard, formalized and centralized processes, involving planned searches, scrupulously sticking with a cycle of gathering, analyzing, estimating and disseminating supposed enriched information."[56] The paradigm did not stop within the IC, either. As Pierre Baumard notes, it was also imported, unchanged, by corporations.[57] However, the range

---

[55] The opinions expressed are those of the author and do not represent those of the National Security Agency, the Department of Defense, or the Office of the Director of National Intelligence.

[56] Philippe Baumard, "From Noticing to Making Sense: Using Intelligence to Develop Strategy," *International Journal of Intelligence and CounterIntelligence*, vol. 7, no. 1 (Spring, 1994), 30. Cited hereafter as Baumard, "From Noticing to Making Sense."

[57] Baumard, "From Noticing to Making Sense," 30.

of phenomena noticed by intelligence professionals has broadened from a focus on largely static issues to encompass highly dynamic topics over the two decades since the end of the Cold War. Intelligence professionals are challenged to stay abreast. A growing professional literature by intelligence practitioners discusses these trends and their implications for advising and warning policymakers.[58]

The literature by practitioners embodies a trust that national intelligence producers can overcome the "inherent" enemies of intelligence to prevent strategic intelligence failure.[59] The disparity between this approach and accepting the inevitability of intelligence failure has grown sharp enough to warrant the identification of separate camps or schools of "skeptics" and "meliorists."[60] As a leading skeptic, Richard Betts charitably plants the hopeful note that in ambiguous situations, "the intelligence officer may perform most usefully by not offering the answer sought by authorities but by forcing questions on them, acting as a Socratic agnostic."[61] However, he completes this thought by declaring, fatalistically, that most leaders will neither appreciate nor accept this approach.

Robert Jervis resurrects a colorful quote from former President Lyndon Johnson, who epitomized the skeptical policymaker:

> Let me tell you about these intelligence guys. When I was growing up in Texas we had a cow named Bessie. I'd go out early and milk her. I'd get her in the stanchion, seat myself and squeeze out a pail of fresh milk. One day I'd worked hard and gotten a full pail of milk, but I wasn't paying attention, and old Bessie swung her s[..]t-smeared tail through the bucket of milk. Now, you know

---

58 The author previously explored this topic in David T. Moore, *Creating Intelligence: Evidence and Inference in the Analysis Process*, MSSI Thesis chaired by Francis J. Hughes (Washington, DC: Joint Military Intelligence College, July 2002) and David T. Moore, *Critical Thinking and Intelligence Analysis*, Occasional Paper Number Fourteen (Washington, DC: National Defense Intelligence College, 2006). Earlier work completed with coauthor Lisa Krizan also included such an examination. See, for example David T. Moore and Lisa Krizan, "Core Competencies for Intelligence Analysis at the National Security Agency," in Russell G. Swenson, ed., *Bringing Intelligence About: Practitioners Reflect on Best Practices* (Washington, DC: Joint Intelligence Military College, 2003), 95-132. Other recent work includes the writings of a number of Intelligence Community practitioners collected by Roger Z. George and James B. Bruce in *Analyzing Intelligence: Origins, Obstacles, and Innovations* (Washington, DC: Georgetown University Press, 2008).

59 Richard K. Betts, *Enemies of Intelligence: Knowledge and Power in American National Security* (New York: Columbia University Press, 2007). Cited hereafter as Betts, *Enemies of Intelligence*.

60 Tamas Meszerics and Levente Littvay, "Pseudo-Wisdom and Intelligence Failures, *International Journal of Intelligence and Counterintelligence* vol. 23, no. 1 (December 2009), 134, 135.

61 Betts, *Enemies of Intelligence*, 51.

that's what these intelligence guys do. You work hard and get a good program or policy going, and they swing a s[..]t-smeared tail through it.[62]

Jervis asserts that policymakers and decisionmakers "need confidence and political support, and honest intelligence unfortunately often diminishes rather than increases these goods by pointing to ambiguities, uncertainties, and the costs and risks of policies."[63] The antagonism is exacerbated when policy is revealed to be flawed and to have ignored intelligence knowledge. For example, in the case of the Bush administration's handling of the Iraq War, intelligence challenges to policy were seen as "being disloyal and furthering its own agenda."[64] Jervis adds that the Bush administration is only the most recent one to exhibit such behavior. He finds that the administrations of Presidents Clinton, Johnson, Kennedy, and Eisenhower also browbeat and ignored intelligence.[65]

Betts, Jervis, and other *skeptics* believe that potential improvements to intelligence processes are limited. Jervis' article on intelligence and policy relations, while it correctly notes the tensions arising from the differing roles of intelligence and policy, over-generalizes the homogeneity of the policy community. It is the author's experience that outside of the highest levels, there are many levels of policymaking that both encourage and welcome the contributions of intelligence. Indeed, some parts of the policy community, beyond the Department of Defense (DoD) where it is the norm to do so, rely strongly on intelligence. Further, disagreements (which Jervis consistently labels conflict) are inherent and typically welcome in the process. Hard questions about the accuracy of judgments must be asked. If we are doomed to such "disagreements," then it is a doom we should be eager to embrace.[66]

The other perspective is that of the *meliorists*—those who feel intelligence processes can be improved. The present authors reside in this camp, preferring to believe that the application of well-informed, mindful expertise, as developed in the present work, can bring positive and substantive value to the fulfillment of the IC's obligations.

---

[62] President Lyndon Johnson quoted by Robert Jervis, "Why Intelligence and Policymakers Clash," *Political Science Quarterly*, vol. 125, no. 2 (Summer 2010), 185. Cited hereafter as Jervis, "Why Intelligence and Policymakers Clash."

[63] Jervis, "Why Intelligence and Policymakers Clash," 187.

[64] Jervis, "Why Intelligence and Policymakers Clash," 190.

[65] Jervis, "Why Intelligence and Policymakers Clash," 190.

[66] Jervis, "Why Intelligence and Policymakers Clash," 204.

Much of the community and its supporting contractors have adopted the meliorist position.[67] As a result, intense attention within and outside the IC has focused on the means by which pertinent phenomena are to be understood. So-called intelligence "analytic" methods are being unshelved or developed and taught to novice and experienced intelligence professionals alike. However, less fully considered are the appropriateness and validity of these methods as well as the underlying assumptions they enshrine. Even less well understood is what happens when specific methods are combined and how those combinations may be made. Several ways exist to characterize these methods in terms of their purpose. However, to date, there is no readily available way to characterize methodological appropriateness or effectiveness, nor the limitations of individual methods. We also lack sound guidance on the use of combined methodologies, despite some recent, promising literature.[68]

A Roadmap

Before these deficiencies can be remedied, however, we need to reframe the way in which intelligence is created. Such a re-conceptualization involves critically examining what intelligence practitioners actually do, and why. The examination demands methodological rigor with particular attention to how we might ensure the validity of our approach to the work of intelligence. If the examination indicates that the existing paradigm for intelligence creation is inadequate, then a revolutionary shift in IC habits will be justified.

Despite the existence of legislative mandates for change, the intelligence-creation process remains largely a product of Cold War-era institutions

---

[67] It should be observed that it is in the interest of IC contractors to adopt this position. As they lobby IC leadership, their sales pitch rests on the idea that their products are best suited for "fixing" the IC's problems. A contractor who is (honestly) skeptical of the possibility that intelligence can be improved is thus likely to see little IC business.

[68] The human species likes to organize knowledge and intelligence professionals are no exception. The list of proposed organizational strategies or taxonomies for intelligence analysis is growing: Morgan Jones developed one such system fifteen years ago in conjunction with his book, *The Thinker's Toolkit: 14 Powerful Methods for Problem Solving* (New York, NY: Random House, Inc., 1995). Babette Bensoussan and Craig Fleisher include taxonomic elements in their catalog of competitive intelligence methods. See for instance, Craig S. Fleisher and Babette Bensoussan, *Strategic and Competitive Analysis: Methods and Techniques for Analyzing Business Competition* (Upper Saddle River, NJ: Prentice Hall, 2002) and *Business and Competitive Analysis: Effective Application of New and Classic Methods* (Upper Saddle River, NJ: FT Press, 2007). The faculty of Mercyhurst College's Institute for Intelligence Studies has issued a taxonomy of methods. Richards Heuer, Jr. developed one on a subcontract to support work performed by Least Squares Software under a contract to IARPA. Another former IC practitioner, Randy Pherson, developed a taxonomy for use in his training programs and subsequently combined his taxonomy with that of Heuer. See Richards J. Heuer, Jr. and Randolph H. Pherson, *Structured Analytic Techniques for Intelligence* (Washington, DC: CQ Press, 2010). To date, none of the taxonomies use cases nor do they ontologize the techniques, that is, show how they are interrelated in process.

and thinking, using the same cognitive frameworks that have been employed for decades. Some argue that what worked in the past is still appropriate. However, as numerous executive and legislative reports confirm, intelligence targets have in fact evolved: adversaries' goals have changed, and their methods have evolved, even if the threats they pose seem very familiar. In sum, the old national intelligence paradigm is woefully out of date.

What is needed now is a discussion of how intelligence can be adjusted to deal with its "traditional" issues as well as new and evolving ones. This book begins that discussion by challenging the standard view of how intelligence professionals do their work. First, as will become clear, intelligence professionals ought *not* be characterized as "analysts." The term is imprecise and inaccurate. If real improvement in intelligence practice is to occur, precision and accuracy in terminology, and thereby in how we think about what we do, are essential first steps.

Intelligence issues are not the same as the issues framed separately by policymakers. To partner successfully with policymakers, intelligence professionals must consider issues from multiple perspectives. This is the role of sensemaking. Yes, the sensemaking process includes "analysis" or attacking issues by "taking them apart." The process also includes synthesis—putting the pieces back together; interpretation—making sense of what the evidence means; and communication—sharing the findings with interested consumers. Essential to these processes is another, that of sound planning or "design."[69] While it could be said that this is what intelligence analysts do, such a statement is epistemologically false. Strictly speaking, intelligence analysts only take issues apart.

So what? Why should we be concerned with a matter of semantics? In short, because the terms we use within the Intelligence Community shape and reflect our practice. If we are to change the culture of intelligence, and be changed by it, our practice of intelligence must also change. New language encourages a new paradigm, and paradigm shifts are revolutionary, not evolutionary. Such a revolution in intelligence is implied in the reform legislation arising out of the 9/11 attacks and the failures to accurately assess the state of Saddam Hussein's programs of weapons of mass destruction.

---

[69] Here, the term "design" mirrors the reflective re-conceptualization of the operational planning process being put in place, as noted earlier, by the School of Advanced Military Studies at Fort Leavenworth.

Kent's Imperative[70]

When much of the tradecraft of intelligence was put in place sixty or more years ago, the dominant framework was that of the historian as scientist. The primary intellectual framework for Cold War intelligence at the national level grew from Sherman Kent's seminal work, *Strategic Intelligence for American World Policy*.[71] Kent's legacy remains active in the National Intelligence Council and the Community at large.[72] Although decision theory and other social science thinking began to influence the creation of intelligence in the 1960s and 1970s, these inputs languished until the reform efforts of recent years. More recently, advances in cognitive science, anthropology, decision theory, knowledge theory, and methods and operations research have brought us to the brink of informed, mindful intelligence sensemaking.

Sherman Kent argues that in creating predictive intelligence about its adversaries "the United States should know two things. These are: (1)…*strategic stature*, (2)…*specific vulnerabilities*."[73] These objectives focus on capabilities and draw heavily from the "descriptive and reportorial elements" of intelligence for basic data.[74] In this way, knowledge about what an adversary ought to do is created. The method by which this is accomplished, according to Kent, is "the one which students reared in the Western tradition have found to be best adapted to the search for truth. It is the classical method of the natural sciences."[75] It involves advice from experts but sees as superfluous to these experts the use of designated red teams—which Kent considers "a new high in human fatuity."[76] If estimates developed from expert judgments are erroneous, he sees the remedy simply in getting more and better information to shed more light on foreign decisionmaking.[77]

---

70 This phrase was originally adopted by an anonymous blogger as the name for a web log of musings on intelligence that ran from 1 January 2006 to 15 October 2008. URL: <http://kentsimperative.blogspot.com>, accessed 6 October, 2010.

71 Kent, *Strategic Intelligence*.

72 Anthony Olcott, "Revisiting the Legacy: Sherman Kent, Willmoore Kendall, and George Pettee—Strategic Intelligence in the Digital Age, *Studies in Intelligence*, vol. 53, no. 2 (June 2009): 21-32.

73 Kent, *Strategic Intelligence*, 40.

74 Kent, *Strategic Intelligence*, 56.

75 Sherman Kent, "Cuban Missile Crisis: A Crucial Estimate Relived," *Studies in Intelligence*, vol. 8, no. 2 (1964), reprint, URL: <https://www.cia.gov/library/center-for-the-study-of-intelligence/kent-csi/vol8no2/pdf/v08i2a01p.pdf>, accessed 27 May 2010, 113. Cited hereafter as Kent, "Crucial Estimate."

76 Kent, "Crucial Estimate," 118.

77 Kent, "Crucial Estimate," 119.

The Failure of an Analytic Paradigm…

Kent's preference for gathering and disaggregating more and more data to find answers fails today in the face of information volume, velocity, and volatility. Marshaling and disaggregating ever more data does not equate to contextual understanding. Further, the assumption that larger pipes to collect data and larger arrays to store it will then allow us to uncover the hidden, clarifying nuggets, is misleading.

Consider what actually happens when intelligence professionals look for an answer to a problem or question. They do not just disaggregate data. Instead, people inquisitively (and selectively) interpret patterns by comparing observed, newly emergent phenomena to what they already "understand." They make sense of phenomena by asking questions; foraging for information; marshaling it into evidence; analyzing, synthesizing, and interpreting that evidence, and communicating their evidence-based understanding of issues to others. Something makes sense because, based on their experience, its pattern is similar to something they previously have seen and that made sense to them. They may even employ a new, self-generated pattern based on previously learned and remembered patterns if they do not get a good match to an ostensible pattern.[78]

Doing so accurately requires making judgments that correlate, according to Air Force thinker William Brei, to the "external world, as it actually exists, regardless of [one's] desires."[79] In other words, one must be able to convincingly correlate ostensible patterns to the data or information for which one is attempting to "make sense." This is not always possible, especially if the phenomenon or issue is broad, novel, or poorly understood; that is, not easily subject to confirmation by universal human sensory apparatii. Brei invokes Ayn Rand on this point:

> To define the meaning of the color "blue," for instance, one must point to some blue objects to signify, in effect: "I mean this."…To define "existence," one would have to sweep one's arms around and say: "I mean this."[80]

---

[78] Dr. David Snowden, conversation with the author, 22 January 2008. Snowden, formerly director of IBM's Institute for Knowledge Management, writes and teaches on complexity and how organizations can leverage it to their advantage. Cited hereafter as Snowden, conversation.

[79] Capt William S. Brei, USAF, *Getting intelligence Right: The Power of Logical Procedure*, Occasional Paper Number Two (Washington, DC: Joint Military Intelligence College, 1996), 17. Cited hereafter as Brei, *Logical Procedure*.

[80] Ayn Rand, *Introduction to Objectivist Epistemology* (New York, NY: Mentor, 1979), 53. Referenced in Brei, *Logical Procedure*, 18.

The social, economic, and political relationships that characterize the govern-ment intelligence milieu mean that severe uncertainty will often remain part of the practitioner's conclusions; telling patterns are at best elusive.

…And a Remedy in Sensemaking

For practitioners to create intelligence knowledge—even with an acknowledged degree of uncertainty—therefore requires much more than mere "analysis." One alternative framework is embodied in the concept of sensemaking. Sensemaking begins with a mindful planning and question-ing that leads to foraging for answers. It is true that along the way the result-ing relevant assemblage of information—or evidence—is disaggregated into its constituent elements. However, it is also synthesized or combined to form a theory or systematic interpretation of the issue that subsequently must be explained, and convincingly. Throughout sensemaking, a continuous assess-ment is demanded of both the processes by which the intelligence is cre-ated and of the intelligence knowledge itself.[81] Mindfulness—as discussed above in the Preface—coupled with a critical thinking-based approach, pro-vide the vigilance, awareness, and self-reflection needed to assess an issue rigorously. This is a central point: Intelligence does not exist in a vacuum. It must contribute to the understanding of an issue by informing the con-cerned parties of a perspective or information they did not already know. Ultimately, if no one is concerned about the knowledge sensemakers create, it is not intelligence.[82]

Karl Weick sees sensemaking as a multiple-step process by which someone goes from becoming aware of "something, in an ongoing flow of events, something in the form of a surprise, a discrepant set of cues, [or] something that does not fit," to a useful understanding of the phenome-non.[83] This definition, which allows for a focus on the social and political environments in which sensemaking takes place, applies to the concept as developed in the present book.

---

[81] For a further discussion of assessing both the process and product of intelligence, see David T. Moore and Lisa Krizan, "Core Competencies for Intelligence Analysis at the National Security Agency," in *Bringing Intelligence About: Practitioners Reflect on Best Practices*, Russell Swenson, ed. (2004), 95-131; and David T. Moore, Lisa Krizan, and Elizabeth J. Moore, "Evalu-ating Intelligence: A Competency-Based Approach," in the *International Journal of Intelligence and CounterIntelligence*, vol. 18, no. 2 (Summer 2005): 204-220. Cited hereafter as Moore, Krizan, Moore, "Evaluating."

[82] See Lois Foreman-Wernet, "Rethinking Communication: Introducing the Sense-Making Methodology," in Brenda Dervin, Lois Foreman-Wernet, and Eric Lauterbach, *Sense-Making Methodology Reader: Selected Writings of Brenda Dervin* (Cresskill, NJ: Hampton Press, Inc., 2003), 1-10. The authors consider communicating an essential part of sensemaking.

[83] Karl E. Weick, *Sensemaking in Organizations* (Thousand Oaks, CA: Sage Publications, Inc., 1995), 2. Cited hereafter as Weick, *Sensemaking*.

Building on Weick's definition and work he did with Kathleen Sutcliffe, Warren Fishbein and Gregory Treverton observe that sensemaking is about anticipating uncertainty as opposed to reacting to it.[84] This means that the processes of sensemaking, and particularly collaborative sensemaking, are never satisfied with the status quo. Rather, sensemaking institutions constantly admit and raise doubts about what they believe. Because threats—as typified by many 21st century issues—can emerge "at any time, anywhere, and in a variety of forms, analysts need to think more in terms of a broad mental readiness to perceive early warning signs."[85]

IARPA, the Intelligence Advanced Research Projects Activity, employs a definition of sensemaking that is complementary to that developed here.[86] They propose that sensemaking is "a core human cognitive ability [that] underlies intelligence analysts' ability to recognize and explain relationships among sparse and ambiguous data."[87] This book accepts that perspective and develops the psychological, behavioral, and social levels of sensemaking as they apply to intelligence creation. By contrast, IARPA's own program on sensemaking seeks to build upon advances in computational cognitive neuroscience that reveal "the underlying neuro-cognitive mechanisms of sensemaking."[88]

As characterized by Peter Pirolli, the process of sensemaking is highly iterative, involving a foraging loop and a sensemaking loop.[89] In the former the sensemaker seeks information, "searching and filtering it," while in the

---

[84] Fishbein and Treverton, "Making Sense," 17. See also, Karl E. Weick and Kathleen M. Sutcliffe, *Managing the Unexpected: Assuring High Performance in an Age of Complexity* (San Francisco, CA: Jossey-Bass, 2001).

[85] Fishbein and Treverton, "Making Sense," 18.

[86] See IARPA Broad-Agency Announcement IARPA-BAA-10-04, *Integrated Cognitive-Neuroscience Architectures for Understanding Sensemaking* (ICArUS) Program, 1 April 2010. URL: <http://www.iarpa.gov/solicitations_icarus.html>, accessed 1 June 2010. Cited hereafter as IARPA, BAA-10-04.

[87] IARPA, BAA-10-04, 4.

[88] IARPA, BAA-10-04, 4. On the emerging discipline of cognitive neuroscience, see The 4th Computational Cognitive Neuroscience Conference, URL: <http://ccnconference.org/>, accessed 7 June 2010.

[89] Peter Pirolli and Stuart Card, "The Sensemaking Process and Leverage Points for Analyst Technology as Identified Through Cognitive Task Analysis," *2005 International Conference on Intelligence Analysis*, McLean, VA, 2-6 May, 2005, URL: <https://analysis.mitre.org/proceedings/Final_Papers_Files/206_Camera_Ready_Paper.pdf>, accessed 18 August 2010. Cited hereafter as Pirolli and Card, "Sensemaking." Pirolli and Card's work here builds on earlier work. See particularly Peter Pirolli and Stuart Card, "Information Foraging," *Psychological Review*, vol. 106, no. 4 (October 1999): 643-675; and Dennis M. Russell, Mark J. Stefik, Peter Pirolli, and Stuart Card, "The Cost Structure of Sensemaking," paper presented at the INTERCHI '93 Conference on Human Factors in Computing Systems, Amsterdam, NL, 24-25 April 1993, URL: <http://www2.parc.com/istl/groups/uir/publications/items/UIR-1993-10-Russell.pdf>, accessed 18 August 2010.

latter an iteratively developed mental model or schema is developed "that best fits the evidence."[90] While the overall flow is "from raw information to reportable results," top-down and bottom-up processes act in concert to reframe issues: information either does or does not fit the hypotheses being considered; hypotheses are refuted or refined, and the larger issue and its context are also reframed, as it comes to be more thoroughly understood.[91] How this can occur within the context of intelligence creation is developed in the following chapters.

To sum up, this book argues that intelligence built around a model of disaggregation as it originated with and developed under Kent, and is still largely practiced today, is at best insufficient. A paradigm based on the concept of sensemaking and employing insights from other knowledge-creation disciplines provides a more appropriate means of skillfully creating intelligence. This book draws a general picture of 21st Century intelligence under a revolutionary paradigm, although it does not explain how all its contours can be fleshed out. We believe that intelligence could be a true profession and moving toward that goal is our desire.[92]

---

[90] Pirolli and Card, "Sensemaking," 3.

[91] Pirolli and Card, "Sensemaking," 3.

[92] In 1960, the Office of Personnel Management (OPM) defined the role of an "intelligence research specialist" as administrative work, not professional work. As it stands, professionals with "state of the discipline knowledge" are by definition excluded from intelligence work. See United States Office of Personnel Management, Workforce Compensation and Performance Service, *Introduction to the Position Classification Standards*, TS-107 August 1991, URL: <http://www.opm.gov/fedclass/gshbkocc.pdf>, accessed 11 December 2009; and United States Office of Personnel Management, *Position Classification Standard for Intelligence Series*, GS-0132 TS-28 June 1960, TS-27 April 1960, URL: <http://www.opm.gov/fedclass/html/gsseries.asp>, last accessed 11 December 2009.

CHAPTER 2
The Failure of "Normal Intelligence"

Intelligence Challenges

Our understanding of everyday phenomena is confounded by everyday strategies employed to mitigate cognitive dissonance, a stressful condition arising when reality clashes with one's perceptions. Two broad strategies, selective exposure and selective perception, can prevent dissonance, but at the expense of sound, mindful reasoning. Through the former, we limit the evidence to that which agrees with or otherwise supports our positions; in the latter, we interpret what we experience in terms of our pre-existing worldview.[93] Examples abound, for these strategies are inherent to the human species. A non-intelligence example appeared in a 2008 broadcast of National Public Radio's *This American Life*, in the story "What Part of 'Bomb' Don't You Understand?" In the broadcast, BBC commentator Jon Ronson juxtaposes the stories of London subway bomb survivor Rachel North and conspiracy theorists who claimed the entire event was fabricated by the British government. The conspiracy theorists even claimed that Rachel North was not an actual person despite her well-documented reality.[94]

Instructive accounts of 9/11 conspiracies and others appear in Farhad Manjoo's *True Enough: Learning to Live in a Post-Fact Society*. One example, related to 9/11, is a belief by some that the second plane that struck the World Trade Center first launched a missile into the building.[95] In both of the above cases, the conspiracy theorists only selected evidence consistent with their conspiracy world-views. What of the situation within intelligence sense-making circles? If the phenomena of both selective exposure and selective

---

[93] For a highly readable discussion of cognitive dissonance, see Carol Tavris and Elliot Aronson, *Mistakes Were Made (But Not by Me): Why We Justify Foolish Beliefs, Bad Decisions, and Hurtful Acts* (Orlando, FL: Harcourt, 2007). For more on selective exposure and selective perception see Farhad Manjoo, *True Enough: Learning to Live in a Post-Fact Society* (Hoboken, NJ: John Wiley & Sons, 2008), especially Chapters 2 and 3. Cited hereafter as Manjoo, *True Enough*.

[94] Jon Ronson, "What Part of 'Bomb' Don't You Understand?" *This American Life*, episode 338, 3 August 2008, URL: <http://www.thislife.org/Radio_Episode.aspx?episode=338>, accessed 4 August 2008.

[95] Manjoo, *True Enough*, 74-80. Its proponents do not adequately explain the logic of such a claim. One key question is "What purpose would such a missile serve?" The proponents of this notion fail to answer this and a number of other important questions.

perception are common, then their effects on intelligence professionals deserve fuller study. It is not understood, for example, how much selective use of evidence typically occurs in the creation of intelligence assessments and estimates.

However, intelligence professionals cannot afford to consider only information that conforms to their own pre-existing worldview or agreed-upon, collective perspective. It is likely that selective exposure and selective perception contributed to the "failures of imagination" noted by the authors of the 9/11 Commission Report: U.S. intelligence professionals and policy-makers in two U.S. administrations failed to make sense of the events leading up to the 11 September 2001 disaster.[96] Failure of imagination was so pervasive a factor that the 9/11 Commission Report found that even those who were oriented toward the threat, such as Richard Clarke, failed to adequately imagine the events of that tragic day.[97]

Errors and Failures

A first step in understanding the lack of sensemaking prior to the 11 September 2001 attacks and other similar events is to understand the differences between "intelligence error" and "intelligence failure." Anthropologist Rob Johnston defines intelligence error in terms of "factual inaccuracies in analysis resulting from poor or missing data."[98] Conversely, intelligence failures are "systemic organizational surprise resulting from incorrect, missing, discarded, or inadequate hypotheses."[99] Thus, the term "failure of imagination" makes sense as a synonym for intelligence failure, where members of an intelligence creating organization fail to imagine in advance the essential outlines of an incident that subsequently occurs.

Additionally, one must consider policy failures. Characterized simply, this failure is seen as the failure to act on intelligence received, and it occurs at many levels. Bruce Berkowitz argues that these errors arise when policymakers "blindside themselves by how they perceive intelligence, by

---

[96] National Commission on Terrorist Attacks Upon the United States, *The 9/11 Commission Report* (Washington, DC: Government Printing Office, 2004), 339-360. Cited hereafter as *The 9/11 Report*.

[97] *The 9/11 Report*, 344. Richard Clarke's memo to Condoleezza Rice warning of the aftermath of such an attack understates the consequences. Clarke considers hundreds dead, not thousands. See Dan Eggen and Walter Pincus, "Ex-Aide Recounts Terror Warnings: Clarke Says Bush Didn't Consider Al Qaeda Threat a Priority Before 9/11," *The Washington Post*, 25 March 2004, A01.

[98] Rob Johnston, *Analytic Culture in the U.S. Intelligence Community: An Ethnographic Study* (Washington, DC: Center for the Study of Intelligence, 2005), 6. Cited hereafter as Johnston, *Analytic Culture*.

[99] Johnston, *Analytic Culture*, 6.

the mental hurdles intelligence must surmount before it can change their perceptions, and in the constraints that limit their ability to act on information…deep down, officials seem to want intelligence to make decisions for them, when in reality, it rarely can."[100] Thus Admiral Husband Kimmel's failure to anticipate the Japanese attack on Pearl Harbor can be attributed (in part) to an inability to overcome a preexisting view about the (in)vulnerability of U.S. forces in Hawaii and particularly at the Pacific Fleet headquarters of which he was in charge.[101] This was exacerbated presumably by a degree of uncertainty in the intelligence. In order to take preventative measures, Kimmel had to *act* based on intelligence reporting, not *react* to it. In considering intelligence-based policy failures, one must consider that despite these explanations, it is the job of intelligence to make sure policy "gets it" and therefore intelligence (or at least its presenters) must also share in the blame. Thus the briefer of a senior policymaker bears a degree of responsibility if the message is not effectively transmitted and acted upon. That this is a difficult task at best must be noted. Policymakers—as is widely noted—often have their own agendas. Would they use intelligence to further them? Fen Osler Hampson argues this was the case in (at least) three separate crises involving Cuba (1962, 1973, and 1979).[102]

Intelligence failures, policy failures and their resulting crises are a regularly recurring theme in U.S. intelligence and policy from at least the mid-20th Century, and likely earlier, to the present. A list of such failures includes:

- Japan's attack on Pearl Harbor (intelligence failure, policy failure).[103]

---

100 Bruce Berkowitz, "U.S. Intelligence Estimates of Soviet Collapse: Reality and Perception," in Francis Fukuyama, ed., *Blindside: How to Anticipate Forcing Events and Wild Cards in Global Politics* (Washington, DC: Brookings Institution Press, 2007), 30. Cited hereafter as Berkowitz, "Soviet Collapse."

101 See Francis Fukuyama, "The Challenges of Uncertainty: An Introduction," in Francis Fukuyama, ed., *Blindside: How to Anticipate Forcing Events and Wild Cards in Global Politics* (Washington, DC: Brookings Institution Press, 2007), 1.

102 Fen Osler Hampson, "The Divided Decision-Maker: American Domestic Politics and the Cuban Crises," *International Security*, vol. 9, no. 3 (Winter, 1984-1985), 130. Cited hereafter as Hampson, "The Divided Decision-Maker." The ODNI issued *Intelligence Community Directive Number 203: Analytic Standards* (June 21, 2007), URL: <http://www.fas.org/irp/dni/icd/icd-203.pdf>, accessed 16 October 2010. The second standard is "Independent of Political Considerations" (p 2).

103 This failure consisted of intelligence policy and errors, as well as intelligence and policy failures. However, hindsight consideration of the events clouds the fact that it was not clear until just before the Japanese attack that an attack was indeed likely. That it was a possibility was, however, discussed. At Pearl Harbor, however, no measures were taken to mitigate the impact of such an attack: Aircraft remained tightly packed on landing areas and ships lacked torpedo nets. The result was strategic surprise at all levels. The best account remains Roberta Wohlstetter, *Pearl Harbor: Warning and Decision* (Stanford, CA: Stanford University Press, 1962).

- North Korea's invasion of the South and China's involvement in the subsequent war (policy failure, intelligence failure);[104]
- The Soviet Union's deployment of IRBM and MRBM nuclear missiles in Cuba (intelligence failure);[105]
- The Vietnamese Tet Offensive (policy failure, intelligence error);[106]
- The fall of the Shah of Iran (intelligence failure);[107]
- The Soviet Union's invasion of Afghanistan (intelligence failure);[108]

---

[104] There is controversy as to whether the invasion of South Korea by North Korea was predicted in advance. Common wisdom considers that no prediction was made. However, Major General Charles A. Willoughby's Korean Liaison Office (KLO) did predict the likelihood of invasion in the spring of 1950 (Kenneth J. Campbell, "Major General Charles A. Willoughby: A Mixed Performance," unpublished paper, URL: <http://intellit.muskingum.edu/wwii_folder/wwiifepac_folder/wwiifepacwilloughby.html>, accessed 5 January 2010). However, according to D. Clayton James, Willoughby had so alienated himself from the CIA and the State Department's intelligence bureau that his warnings apparently were ignored by the civilian intelligence components. The Army's G-2 apparently ignored his warning as well, possibly because the United States Military Advisory Group to the Republic of Korea, and not the KLO, was tasked with such estimating. See also, D. Clayton James, *Years of MacArthur, 1945-1964* (Boston, MA: Houghton Mifflin, 1985), 416. The fact that warnings were issued but ignored adds the policy failure component.

[105] DCI John McCone was at first skeptical and then was convinced that there were missiles in Cuba. Refugee reports during the summer of 1962 suggested that the missiles were based in Cuba. However, the Office of National Estimates under the leadership of Sherman Kent predicted the Soviets would not deploy missiles in Cuba. U-2 overflights did not detect the presence of the missiles until mid October, although they had earlier detected the presence of defensive Surface to Air Missiles. Lacking confirmatory technical corroboration, the State Department and the White House were not willing to act based on refugee reports (which had been received since 1960 and up to this point apparently had been incorrect). See Linda K. Miller and Mary McAuliffe, "The Cuban Missile Crisis," *Magazine of History*, vol.8 (Winter 1994), URL: <http://www.oah.org/pubs/magazine/coldwar/miller.html>, accessed 5 January 2010.

[106] According to Harold Ford, despite analyses to the contrary, wishful thinking by key policymakers and their political pressure on the Intelligence Community to concur led to an underestimation of the Viet Cong and North Vietnam's military capabilities. Such a view precluded the possibility of a Tet-like offensive. Thus systemic surprise from a policy failure is associated with the Tet Offensive. See Harold Ford, *CIA and Vietnam Policymakers: Three Episodes 1962-1968* (Washington, DC: Center for the Study of Intelligence, 1998), 85-138.

[107] In writing about failure, Bruce Berkowitz quotes the CIA August 1978 assessment that "Iran is not in a revolutionary or even pre-revolutionary state." See Bruce Berkowitz, "U.S. Intelligence Estimates of Soviet Collapse: Reality and Perception," in Francis Fukuyama, ed., *Blindside: How to Anticipate Forcing Events and Wild Cards in Global Politics* (Washington, DC: Brookings Institution Press, 2007), XX. Cited hereafter as Berkowitz, "Soviet Collapse."

[108] Doug MacEachin writes that the failure to predict the Soviet invasion of Afghanistan "illustrates probably the most recurrent trap for analysts...One part of it might be called the 'model cage.' Once having constructed an intellectual model of how the variables are likely to play out, each new piece of information is weighed in accordance with the components of that model. Evidence that does not fit is far more likely to be explained away than used to question the model's validity. In this case, the actions taken (military preparations) were not used to interpret intentions so much as the conclusions about intentions were used to interpret the actions." See Douglas MacEachin, *Predicting the Soviet Invasion of Afghanistan: the Intelligence Community's Record* (Washington, DC: Center for the Study of Intelligence, 2002), URL: <https://www.cia.gov/library/center-for-the-study-of-intelligence/csi-publications/books-and-monographs/predicting-the-soviet-invasion-of-afghanistan-the-intelligence-communitys-record/predicting-the-soviet-invasion-of-afghanistan-the-intelligence-communitys-record.html>, accessed 5 January 2010.

- The collapse of the Soviet Union (intelligence failure?, policy failure?);[109] and

- Iraq's 1990 invasion of Kuwait (intelligence failure, policy failure).[110]

Similarly, professionals apparently failed to make sense of the precursors of the 25 December 2009 attempt to bring down a U.S. airliner over Detroit. Here again, failures of imagination that accompanied the centrifugal disaggregation of data gathering and evaluation on the "underwear bomber" contributed to the scenario of not performing early cross-checking of no-fly lists and other terrorist-related databases, as well as not accepting the father of a motivated Islamic radical as a credible source.[111]

These intelligence errors and failures have occurred as the IC has continued in the Kent vein of seeking more and better data, but without framing the issues in a way that allows the national intelligence process to use its special capabilities to apply deductive or even abductive logic to foraging for, marshaling and evaluating data. At the same time, intelligence oversight reports by Congressional Committees and Special Commissions

---

[109] There is considerable disagreement about whether the Community failed to predict the collapse of the Soviet empire. For a summary of both sides of the issue with references, see Gerald K. Haines and Robert E. Leggett, ed., "Introduction and Overview of the Conference Papers" in *Watching The Bear: Essays on CIA's Analysis of the Soviet Union* (Washington, DC: Center for the Study of Intelligence, 2003), URL: <https://www.cia.gov/library/center-for-the-study-of-intelligence/csi-publications/books-and-monographs/watching-the-bear-essays-on-cias-analysis-of-the-soviet-union/index.html>, accessed 5 January 2010. In taking the position that the community *did* warn, Bruce Berkowitz argues the "record suggests that U.S. intelligence provided about as good a product as one could reasonably expect...[It] stipulated a set of conditions...and it notified top U.S. leaders when these conditions were met." If not an intelligence failure, then what? Berkowitz offers clues to this question in noting that leaders blindside themselves by failing to understand intelligence—in this case their failing to "get it" about the intelligence coming out of the IC leads one to conclude the failure was theirs—a policy failure. Additionally, he notes that popular belief in the failure, an example of selective perception, was also the result of key documents remaining classified, a sort of selective exposure. See Berkowitz, "Soviet Collapse," 29-30, xx.

[110] Don Oberdorfer, writing in *The Washington Post*, notes that policymakers admitted to being "guilty of a kind of mind-set or framework about Iraq." They failed to consider that the Iraqis would go beyond saber rattling to invade Kuwait. Oberdorfer also quotes administration officials as admitting that they did not focus on Iraq because they "didn't have the time." See Don Oberdorfer, "Missed Signals in the Middle East, *The Washington Post Magazine*, 17 March 1991, 40.

[111] The Senate Select Committee on Intelligence released its report on the "underwear bomber" incident on 18 May 2010 (see *http://intelligence.senate.gov/100518/1225report.pdf*). Of 14 points of failure identified, 12 were failures of intelligence process, including interpretation. Failures to correctly interpret source information are a part of the other failures cited above. For example, such failures also occurred in the case of the Cuban Missile crisis where (as we discovered some 40 years later) the refugees were telling us the truth: there were nuclear missiles in Cuba during the summer of 1962; they simply were not strategic missiles. Rather they were tactical, nuclear-tipped cruise missiles. See Raymond L. Garthoff, "US Intelligence in the Cuban Missile Crisis," in James G. Blight and David A. Welch, eds., *Intelligence and the Cuban Missile Crisis* (London, UK: Frank Cass, 1998), 29. Cited hereafter as Garthoff, "US Intelligence."

have repeatedly faulted the community for its lack of imagination in antici-pating at least the grand design of events that, instead, surprised or shocked nearly everyone associated with this community. This unhappy situation suggests that there are flaws in Kent's model of disaggregating significant amounts of data in order to predict specific events. The etymology of "imag-ination"—generating images—reminds us of the contemporary critic of Kent, Willmoore Kendall, who suggested that the job of national intelli-gence is to communicate with decisionmakers in a "holistic" way so as to generate the "pictures [mental models] that they have in their heads of the world to which their decisions relate."[112]

Considering Standard Models

Intelligence failures occur as practitioners employ a "standard model"[113] of intelligence: In it, analysts "separate something into its constitu-ent elements[114] so as to find out their nature, proportion, function, relation-ship, etc.[115] and "produce reports" based on "collected" information and data. There is a definitional presumption that disaggregation will lead to answers. However, this model *incompletely* describes what the intelligence professional does and its underlying presumption about finding answers may be false.

One problem is that in Kent's data-based analytic framework, analysts need to have all the data available so they can be marshaled into a coher-ent account. "Dots"—if they exist at all—can be connected in more than one way.[116] In foresight it is difficult at best to determine which combination and order is valid. Such determinations can be complicated further by the fact that adversaries may change their actions if they suspect we have arrived at a certain conclusion.

An additional problem is that with an increased number of signals there is also an increased level of noise. Which signals, which facts, or which inferences the intelligence professional should consider valid becomes a very important consideration. At best, warning of a pending incident is a prob-lem of assembling and making sense of the details of a specific incident in advance. However, many intelligence problems inherently defy such linear

---

112 Willmoore Kendall, "The Function of Intelligence," *World Politics*, vol. 1, no. 4 (July 1949), 550. Cited hereafter as Kendall, "Function of Intelligence."

113 A standard model is one that is widely accepted to be justified and true. It is—in day-to-day considerations—sufficient and therefore its validity is not questioned.

114 *New Oxford American Dictionary*, Apple Computer Edition, entry under "analysis."

115 Webster's New Universal Unabridged Dictionary, 2nd ed. 1983, entry under "Analysis."

116 Robert Horn prefers to refer to dots as "smudges," suggesting that they are at best imprecise in both existential and contextual frameworks. Robert Horn, conversation with the author, 6 October 2010.

characterization. They are in fact "wicked" problems—a formal designation of a complex issue with myriad linkages. We turn next to an exploration of problem types to see how their nature directs our making sense of them.

Types of Problems

In order to understand "wicked problems," one must first understand the nature of "tame problems."

Tame Problems

In a tame problem there is general agreement as to what or who an adversary is, what the "battlefield area" is, and what an attack is. Such problems, while difficult, exhibit specific characteristics: They are clearly defined and it is obvious when they are solved. Solutions to these problems arise from a limited set of alternatives that can be tested; the correct solution can be objectively assessed. Finally, solving one tame problem can facilitate creating valid solutions to other, similar tame problems.[117]

It is important to note that analysis protocols for tame problems contain little or no room for "emergent" properties. One may not know that the analytic protocol is insufficient until the puzzle has been incorrectly defined, characterized, and solved, if it is in fact solvable. One arrives at one solution that at first appears to have resolved the issue, but in fact, the issue reemerges elsewhere. For example, the implementation of a linear, intelligence-driven solution to crack down on insurgents and their improvised explosive devices (IEDs) in one area may lead to an emergence of IED-caused explosions somewhere else. In such a case, the application of "tame problem protocols" may in fact have been inappropriate—the problem is in fact not tame.

Admittedly, many 21st Century intelligence issues remain puzzles or tame problems. This occurs when the events surrounding the issues have already occurred, appropriate questions are readily identifiable, and answers exist, even if they are difficult to find. For example, in a weapons proliferation puzzle, if we know that missiles have been built, the nature of their warheads and their accuracy "may remain unknown even though they are knowable."[118] Solution is a process of discovery and sensemaking.

Seen in this light, even the attacks on the United States by Al Qaeda on 11 September 2001 could be considered a puzzle or tame problem. Plans

---

117 Jeff Conklin, *Dialogue Mapping: Building Shared Understanding of Wicked Problems* (Hoboken, NJ: John Wiley and Sons, Ltd., 2005), 9.

118 Gregory F. Treverton, *Reshaping National Intelligence for An Age of Information* (Cambridge, UK: Cambridge University Press, 2001), 12. Cited hereafter as Treverton, *Reshaping National Intelligence*.

had been made, necessary skills (flying airplanes) learned, surveillance conducted, targets selected, weapons acquired, and terrorists positioned. The issue faced by intelligence professionals—and where we failed—was to sense and then figure out all (or at least enough) of the pieces before the events of that day occurred. We also had to figure out what the "event" was. The difficulty of doing so at both a theoretical and practical level points out how difficult tame problems can be to solve. As noted, in the end we failed, although some pieces of the puzzle—such as the flying skills necessary—at least had been sensed. In this case, the essence of the puzzle itself— the intention to deliberately fly passenger airplanes into structures in the U.S.—remained unidentified.

Wicked Problems

However, seen in a larger context, are such puzzles truly tame? Or are they components—as Russell Ackoff suggests—of something larger: a mystery in Treverton's terms, or a "mess" according to Ackoff.[119] Treverton's intelligence mysteries defy easy definition. They belong to a class of problems defined by social researchers Horst Rittel and Melvin Webber as "Wicked Problems." In describing this domain Rittel and Webber note that in considering wicked problems and systems there are a great many barriers to sensemaking:

> [Theory] is inadequate for decent forecasting; our intelligence is insufficient to our tasks; plurality of objectives held by pluralities of politics makes it impossible to pursue unitary aims; and so on. The difficulties attached to rationality are tenacious, and we have so far been unable to get untangled from their web. This is partly because the classical paradigm of science and engineering—the paradigm that has underlain modern professionalism—is not applicable to the problems of open societal systems.[120]

The adaptive nature of adversaries makes seemingly tame puzzles wicked, moving them into the realm of "unknown unknowables."

By definition, wicked problems are "incomplete, contradictory, and changing."[121] They do not have single answers and in fact, are never truly

---

119 Treverton, *Reshaping National Intelligence*, 11-13; Russell A. Ackoff, *Redesigning the Future: A Systems Approach to Societal Problems* (New York, NY: John Wiley and Sons, 1974), 11.

120 Horst W. J. Rittel and Melvin M. Webber, "Dilemmas in a General Theory of Planning," *Policy Sciences*, vol. 4 (1973), 157. Cited hereafter as Rittel and Webber, "Dilemmas."

121 Wikipedia, Entry under "Wicked Problem," accessed 28 March 2007. Cited hereafter as Wikipedia, "Wicked Problem."

answered. In the context of intelligence, the sensemaker may never realize a problem has been resolved. This is because "the solution of one of its aspects may reveal or create another, even more complex problem."[122] The emergent complexity of the problem itself, its adaptive nature, efforts at denial and deception by adversarial actors, as well as cognitive frailties on the part of sensemakers, compound the problem, confounding sensemaking, leading in some cases to disastrous courses of action or consequences.

Table 1. Characteristics of Wicked Problems
Wicked problems have no definite formulation.
Wicked problems have no clear end-point.
Solutions to wicked problems are at best good or bad.
Tests of solutions to wicked problems may not demonstrate their validity and may provoke undesired consequences.
Implementing solutions to wicked problems changes the problem.
Sensemakers can never know if they have determined all the solutions to wicked problems.
Each wicked problem is essentially unique.
Every wicked problem is embodied in another one.
How wicked problems are resolved is determined by the means and methods used to make sense of them.
Sensemakers have no right to be wrong.

Source: Derived from Horst W. J. Rittel and Melvin M. Webber, "Dilemmas in a General Theory of Planning," *Policy Sciences* 4 (1973), 155-169.

Rittel and Webber note that all wicked problems share at least 10 characteristics in common (summarized in table 1). Wicked problems so framed allow us to proceed with discussions into their nature. The two men argue, however, that our standard "basis for confronting problems of social policy is bound to fail, because of the nature of these problems." The means that we typically have at hand for cognitive handling of these problems "is not applicable."[123]

---

122 Wikipedia, "Wicked Problem."
123 Rittel and Webber, "Dilemmas," 162.

A Wicked Look at Wicked Problems in Intelligence

Characterizing intelligence issues in terms of their problem type—admittedly somewhat vaguely (in keeping with their nature)—reveals just how prevalent wicked problems are within the domains of intelligence.

Wicked problems have no definite formulation. To Rittel and Webber, "the process of solving the problem is identical with the process of understanding its nature, because there are no criteria for sufficient understanding."[124] In other words, making sense of problems deemed sufficiently complex so as to be considered wicked is equivalent to characterizing them in the first place; the description encompasses all possible solutions.

For example, one wicked problem could be "how best to stem the growth of terrorism in the Middle East." An assumption in considering this problem is that if intelligence professionals can understand what motivates people to become terrorists in the first place, intervention might be possible. Mitigating the creation of new terrorists could aid in reducing both their numbers and by extension, their attacks. Do people become terrorists because they are dissatisfied with what they see as contradictions and hypocrisies in their lives? If so, what then are the specific roots of dissatisfaction and contradiction? One commonly cited reason is a lack of economic opportunity for males within societies. In that light, Rittel and Webber ask, "Where within the…system does the real problem lie? Is it [a] deficiency of the national and regional economies, or is it deficiencies of cognitive and occupational skills within the labor force?"[125] The possible solutions to this problem extend the domain of questions, spreading ever outward.[126]

Our ignoring domains that seem irrelevant (either for practical or political reasons) is a strategy of selective exposure and perception. Admittedly, some domains may be inconsequential. On the other hand, one of the interesting features of complex systems is that small perturbations can produce large impacts. So, a decision to eliminate factors from consideration may result in discarding seemingly inconsequential elements, with as yet unknown but major impacts. Such possible impacts cannot be known in advance, as they are a part of the noise surrounding the issue. Indeed, as Nicholas Taleb notes,

> [Our] track record in predicting those events is dismal; yet by some mechanism called the hindsight bias we think that we understand

---

124 Rittel and Webber, "Dilemmas," 162.
125 Rittel and Webber, "Dilemmas," 161.
126 Such domains include (among others) economics, culture, history, geography, roles of language, religion, or law, singly or in combinations.

them. We have a bad habit of finding "laws" in history (by fitting stories to events and detecting false patterns); we are drivers looking through the rear view mirror while convinced we are looking ahead.[127]

Seemingly unimportant factors also are not considered by sensemakers due to a failure to adequately address assumptions about the issue at hand, as noted in official reviews of recent "intelligence failures." For example, the Senate's report on the prewar assessment of weapons of mass destruction in Iraq specifically notes that analysts' assumptions were not challenged in the creation of the estimate.[128]

Even with the addressal of major assumptions, there remain additional underlying factors that do not get questioned—almost an endless succession of assumptions that must be peeled off the problem much as one peels layers off an onion. There is an added complication that individual layers are not sequential and in fact may lead (to continue the analogy) to other onions or other vegetables, or even fruit. In intelligence, such assumptions are themselves a *mess*: a complex system of interrelated experience, knowledge, and even ignorance that affects reasoning at multiple levels sequentially and simultaneously. There is an old English children's nursery rhyme that neatly characterizes this: "For want of a nail the shoe was lost. For want of a shoe the horse was lost. For want of a horse the rider was lost. For want of a rider the battle was lost. For want of a battle the kingdom was lost. And all for the want of a horseshoe nail."[129] Clearly, such a sequence implies a logical progression, whereas in dealing with wicked problems, the order may be mixed up. Some of the links may even be unknown—either missing or unknowable.

Wicked problems have no clear end-point. With tame and well-structured problems one knows when the solution is reached. In wicked problems this is not so, as Rittel and Webber make clear:

> There are no criteria for sufficient understanding and because there are no ends to the causal chains that link interacting open systems, the would-be planner can always try to do better. Some additional investment of effort might increase the chances of finding a better solution.[130]

[127] Nicholas Nassim Taleb, "Learning to Expect the Unexpected," *Edge*, 19 April 2004, URL: <http://www.edge.org/3rd_culture/taleb04/taleb_indexx.html>, accessed 1 March 2007.

[128] United States Senate, *Report on the U.S. Intelligence Community's Prewar Intelligence Assessments on Iraq*, Select Senate Committee on Intelligence, 108th Congress, 7 July 2004, 18.

[129] Anonymous, *For Want of a Nail Rhyme*, URL: <http://www.rhymes.org.uk/for_want_of_a_nail.htm>, accessed 10 September 2007.

[130] Rittel and Webber, "Dilemmas," 162.

This is not a new consideration. Writing in the 1930s, John Dewey observed that "the 'settlement' of a particular situation by a particular inquiry is no guarantee that *that* settled conclusion will always remain settled. The attainment of settled beliefs is a progressive matter; there is no belief so settled as not to be exposed to further inquiry."[131] Intelligence sensemakers routinely confront this challenge. Reports and assessments often update or revise previous conclusions. Often the previous reporting is consulted *before* the new report is written so that the author can determine the preexisting point of view on the issue. Such consultations at best determine whether the current situation deviates from the norm. Unfortunately, sometimes such consultations lead to the rejection of the new evidence, opening the way to intelligence errors and failures. One goal of an adversary's denial and deception activities is to facilitate rejection of the novel deviation. It was in this way that the possibility of nuclear missiles deployed to Cuba was rejected amid outlandish noise during the summer of 1962, and military exercises along the Suez Canal lulled Israel into a sense of creeping normalcy prior to October 1973.

Solutions to problems may be implemented for "considerations that are external to the problem" itself: problem solvers "run out of time, or money, or patience."[132] In intelligence, sensemakers may only be able to work for a given time on a problem before they have to issue their report. Changes in funding may mean that an effort to understand a phenomenon has to be discontinued. The practicalities of resource limitations force changes in sensemakers' foci. However, this does not mean that the problem does not continue to exist and perhaps, threaten. Rather, an answer has been developed to a distilled problem, communicated, and now other things must be done.

Solutions to wicked problems are at best good or bad. Some problems have true or false, yes or no answers. These are not wicked problems. Wicked problems have no such answers. Differing perspectives applied by different problem solvers, differing sets of assumptions, and differing sources of evidence are several of the factors that lead separate groups to come to different judgments about wicked problems. The impossibility of exhaustively considering all the factors and solutions of the problem also contributes to a multiplicity of solutions. At best these can be ranked as good or bad solutions. In most cases, according to Rittel and Webber, the solutions are expressed as "better or worse" or "satisfying" or "good enough."[133]

---

131 John Dewey, *Logic: The Theory of Inquiry* (New York, NY: Henry Holt, 1938), 8-9. Emphasis (italics) in the original.

132 Rittel and Webber, "Dilemmas," 162.

133 Rittel and Webber, "Dilemmas," 163.

Given, for example, the problem of stemming the growth of terrorism, there is no one simple solution that suffices. Instead, a number of differing solutions exist, depending upon (among other things) the perspectives about the domains at work in the problem. Focusing on the economics surrounding the growth of terrorism leads to different proposed solutions than does focusing on the demographics involved in the issue. Religious considerations or broader cultural considerations also create different solutions. Each of these perspectives in turn optimizes multiple points of view with differing, good and bad solutions. Overlap is possible and even desired. Good solutions encompass multiple domains.

Tests of solutions to wicked problems may not demonstrate their validity and may provoke undesired consequences. Implemented solutions to wicked problems "generate waves of consequences over an extended—virtually an unbounded—period of time."[134] Further, these consequences may themselves prove so undesirable as to negate any and all benefits of the original decision—*and this cannot be determined in advance.* Thus an intelligence-based decision to invade a country's possessions may create circumstances that offset any gains initially won, as the Argentineans discovered in 1982 when they—unwisely in retrospect—seized the British-owned Falkland Islands. From the perspective of the Argentine regime, the initial "victory" was offset when Britain forcibly retook the islands. The Argentine Navy lost a capital ship and many died; the regime lost power and was ultimately ousted. Seen from the perspective of the Argentine people this was, in the long term, of benefit. The government-condoned disappearances (torture and murder) of its foes ceased. Democratic processes were restored. However, things could easily have gone in another direction. One repressive regime could have been replaced by another. None of these outcomes was knowable in advance.

Implementing solutions to wicked problems can change the problem. In intelligence problems, real solutions cannot be practiced; there are no "dry runs." True, sensemakers and their policy-making customers can (and should) consider what *might* happen or the "implications" of the decisions or solutions of the problem at hand. Doing so might increase the likelihood that the decision selected is the best or the less bad of a set of bad alternatives.

Modeling the situation is one common means of assessing the implications of a potential action. However, models must by their very nature limit the factors considered. This raises the question of how one might know in advance if the eliminated factors are in fact significant. Further, modeling or any other means of generating solutions does not guarantee that the selected

---

[134] Rittel and Webber, "Dilemmas," 163.

23

decision is the right decision. Once implemented it cannot be undone. It is noteworthy that at this point "implications" that do arise are actually "consequences" and hopefully they have been considered. But additional consequences and responses to those consequences by the actors in the problem as they respond to the "solution" will change, transform and evolve the problem. Taking down a terrorist's "safe house" may not reduce the threat, but does change where and how the remaining terrorists operate. This applies as well to attempts to reverse decisions. As Rittel and Webber note, "every attempt to reverse a decision or to correct for…undesired consequences poses another set of wicked problems,"[135] as sensemakers and planners involved in the U.S.-led "war on terror" have discovered. Actions, once taken, may mitigate the threat, or may not, which leads to the next facet of wicked problems.

Sensemakers can never know if they have determined all the solutions to wicked problems. They can expect, however, that they almost certainly *have not* determined all the solutions. In developing the range of alternatives within scenarios, two goals predominate: mutual exclusivity and collective exhaustion. In other words, each alternative must preclude the simultaneous possibility of the others, *and* the entire set of known alternatives must be considered. In practical terms, this is much more difficult to achieve than it sounds. Intellectual frameworks and so-called "biases" such as vividness, anchoring, confirmation, and others combine to prevent people from being able to consider all the alternatives. Adding to this is the fact that issues evolve in unpredictable ways. All the solutions simply are not knowable because they lie in the future. This does not justify not trying to completely assess the alternatives but rather provides recognition that some alternatives elude consideration.

Each wicked problem is unique. While it is true that common elements can be found between problems, there remain additional and unique properties of "overriding importance."[136] In other words, wicked problems cannot be characterized into "*classes*…in the sense that principles of solution can be developed to fit *all* members of a class."[137] For example, there are common elements or patterns in proliferation that allow recognition by sensemakers: acquisition of certain materials, construction of facilities, and the like. However, denial and deception—if applied—may obscure these commonalities. Knowing specific details of a weapons development program can be elusive. Another element, the *intentions* of the proliferators or the

---

135 Rittel and Webber, "Dilemmas," 163.
136 Rittel and Webber, "Dilemmas," 164.
137 Rittel and Webber, "Dilemmas," 164.

recipients of the proliferated systems are also of critical importance; they may be unique, and perhaps intractable. What Rittel and Webber have to say about this consideration is germane to intelligence: "Despite seeming similarities...one can never be *certain* that the particulars of a problem do not override its commonalities with other problems already dealt with."[138]

Every wicked problem is embodied in another one. Rittel and Webber describe problems as

[discrepancies] between the state of affairs as it is and the state as it ought to be. The process of resolving the problem starts with the search for causal explanation of the discrepancy. Removal of that cause poses another problem of which the original problem is a "symptom." In turn, it can be considered the symptom of still another, "higher level" problem.[139]

What policies and actions, for example, are necessary to "fix intelligence?" Answering this involves asking what is causing intelligence to fail. One place to start is to consider why analysts are wrong and how intelligence errors lead to intelligence failure.[140] Yet such considerations lead one to consider how consumers may ignore intelligence, and how adversaries may in fact be "more capable" than expected. These in turn lead to what Jeffrey Cooper considers "analytic pathologies" that decrement both individual and corporate efforts to make sense of issues (table 2). Each of Cooper's specific pathologies is furthermore at least partially embodied in the others, giving rise to error-producing systems.[141] For example, Cooper argues that intelligence professionals' pathological focus on both "the 'dots' analogy and the model of 'evidence-based' analysis...understate significantly the need for imagination and curiosity."[142] Related to this is what he calls the myth of "Scientific Methodology." Analysis is *not* [hard] science and is *not* about proof. Rather it is about discovery.[143] These are embodied in the protocols he refers to as the flawed "Tradecraft Culture,"— a guild system of potential sensemakers and their historically unchanging ways of working.[144]

---

138 Rittel and Webber, "Dilemmas," 165. Emphasis in original.

139 Rittel and Webber, "Dilemmas," 165.

140 Johnston develops this distinction. Johnston, *Analytic Culture*, 64-66.

141 See Jeffrey R. Cooper, *Curing Analytic Pathologies: Pathways to Improved Intelligence Analysis* (Washington, DC: Central Intelligence Agency, Center for the Study of Intelligence, 2005). Cited hereafter as Cooper, *Analytic Pathologies*.

142 Cooper, *Analytic Pathologies*, 28.

143 Cooper, *Analytic Pathologies*, 28-29.

144 Cooper, *Analytic Pathologies*, 30.

Table 2. Cooper's Analytic Pathologies
An intelligence account system whereby institutions and individuals "own" issues is inefficient. The return on investment for having accountability is too low when compared to the lack of cooperation, collaboration, and sharing such a system promotes. (30-31)
A cultural "evidence-based scientism" that prevents anticipatory consideration of policy and military intelligence consumers' needs. (31)
An overemphasis on current intelligence to the detriment of the "long view" resulting from a chaotic post-Cold War environment of emergent issues and crises. (32)
A production-oriented model that focuses on collecting, and processing massive quantities of data, and producing routine reports without the capability to adroitly refocus resources for ad hoc reports. (32)
A use of previous judgments as the starting point for all subsequent reporting. A corollary is a belief that "finished intelligence" is anything more than a snapshot in time, that it conveys larger "truth." (33)
A neglect of deep research about issues brought about by short-term tasks. (34)
A neglect of anticipatory intelligence arising from attempts by intelligence sensemakers to emulate the other (current) intelligence sources—such as news networks—on which their consumers also rely. (35)
A loss of "keynote species," mid-level sensemakers with deep domain expertise who create the in-depth assessments that convey what specific issues are all about. (36-37)
An inaccurate focus on results instead of processes leads to a failure to develop, validate, and promulgate methods of intelligence sensemaking. (37)
A security mindset leads to a lack of cooperation between intelligence sensemakers, and domain experts and their knowledge regardless of who and where they are. (39)

Source: Derived from Jeffrey Cooper, *Curing Analytic Pathologies: Pathways to Improved Intelligence Analysis* (Washington, DC: Central Intelligence Agency, Center for the Study of Intelligence, 2005), 30-39.

Rittel and Webber posit "incrementalism" as a not uncommon approach to mitigating these effects. Related to Kuhn's concept of "normal science," incrementalism may actually compound the specific problem further because incrementalism

[advertises] a policy of small steps, in the hope of contributing systematically to overall improvement. If, however, the problem is attacked on too low a level (an increment), then success or resolution may result in making things worse, because it may become more difficult to deal with the higher problems. Marginal improvement does not guarantee overall improvement.[145]

In considering Cooper's pathologies, fixing selective individual pathologies—insofar as possible—produces organizational changes that may inhibit further "fixes" by making them monetarily or organizationally too expensive. For instance, "solving" technological problems associated with Cooper's identified pathologies imposes infrastructures that may inhibit other necessary transformations such as developing a more agile workforce.

How wicked problems are resolved is determined by the means and methods used to make sense of them. In other words, how problems are perceived determines the kinds of solutions that are proposed. Point of view becomes essential in defining what a problem is and how it is to be resolved. Complex, wicked problems (as well as many "tame" ones) cannot be defined from one point of view. Defining the causes of terrorism is a case in point. As considered by the participants at the 2005 International Summit on Democracy, Terrorism, and Security, sponsored by the *Club de Madrid*, terrorism's causes lie in five broad domains: psychology, politics, economics, religion, and culture. Yet, as Martha Crenshaw notes in the conference report on the causes of terrorism, such considerations may be invalid:

Explaining terrorism in terms of background conditions (social, economic, demographic, political, or cultural) is insufficient at best, and wrong at worst. Focusing exclusively on underlying structures provides little predictive capacity. "Root causes" may, in fact, influence the subsequent trajectory of terrorism more than its onset since they determine the extent of social support for violence by justifying grievances.[146]

---

[145] Rittel and Webber, "Dilemmas," 165.

[146] Martha Crenshaw, "Political Explanations," in Kim Campbell, ed., *Addressing the Causes of Terrorism: The Club de Madrid Series on Democracy*, vol. 1 (Madrid, SP: Club de Madrid, 2005), 13.

Yet these considerations are used to develop "solutions" to terrorism. Methods that consider and "solve" underlying economic or demographic issues, we may discover, only partly explain the phenomena.

How Are Wicked Problems Disruptive?

Disruption, as developed by Clayton Christensen, emerges from technologies that, while they may under-perform established technologies, open new markets and change the ways people do things. Enlarging the definition, disruptive intelligence problems threaten to change the way people interact. They proffer or impose new paradigms—both "good" and "bad"—for non-governments and governments alike. The disruption occurs because the incumbent is doing the *most rational thing it can do* given its circumstances. Doing the right thing generates the opportunity for disruption. For example, among the disruptions arising from a pandemic could be an easing of population pressures (if enough people die). This could lead to a freeing-up of energy resources for the survivors. Alternately, such a population die-off might cause a breakdown of societal infrastructures leading to riots and chaos. Both outcomes (and others)—as seen in anticipation—share likelihoods: all are likely. None of them can be reliably calculated (and therefore predicted) with any certainty.

See Clayton Christensen, *The Innovator's Dilemma* (Cambridge, MA: Harvard University Press, 1997).

Sensemakers have no right to be wrong. One of the first things many visitors to the CIA first see is the aphorism, "Ye shall know the truth, and the truth shall set you free."[147] As retired CIA veteran Ray McGovern comments,

> [T]he primary function of the Central Intelligence Agency is to seek the truth regarding what is going on abroad and be able to report that truth without fear or favor. In other words, the CIA at its best is the one place in Washington that a President can turn to for an unvarnished truthful answer to a delicate policy problem.[148]

This aphorism may have validity in the domain of tame problems where the truth is known or knowable. However, it has much less (if any) validity in the world of wicked problems where many truths can coexist.

---

[147] John, 8:32 (King James Version).

[148] Will Pitt, "Interview: 27-Year CIA Veteran," *Truthout*, 26 June 2003, URL: <http://www.truthout.org/docs_03/ 062603B.shtml>, accessed 12 March 2007.

Depending on the point of view expressed the context can be simultaneously true and contradictory, and may in fact be unknowable.

The goal of assessing wicked problems may be to "improve some characteristics of the world where people live. Planners are liable for the consequences of the actions they generate; the effects can matter a great deal to those people that are touched by those actions."[149] Intelligence error, regardless of what causes it, is considered intolerable as U.S. policymakers and Intelligence Community sensemakers most recently discovered with their inaccurate estimate on the state of Iraq's programs to develop WMD. Earlier policymakers and sensemakers faced similar situations regarding the intentions of Japan, North Korea, China, Cuba, the Soviet Union (repeatedly), and India. In other words, restating part of Rittel and Webber's quote above yields this guideline: "*intelligence professionals* are liable for the consequences of the intelligence they generate."

An Intelligence Example: Pandemics as Wicked Problems

One of the threats faced by intelligence organizations and their professionals is that of an emergent global pandemic. What kind of a threat is a pandemic? Is it a tame or wicked problem, or something in between? Such considerations matter because they define what approaches are suitable for alleviating or mitigating the threats to national security that pandemics pose.

Historically, pandemic infectious diseases disrupted societies over wide regions of the world. Of these, the bubonic plague pandemic of the mid 14th Century, also called the "Black Death," is perhaps the best known. By killing off approximately one-third of Europe's population, it is credited with ending serfdom in most of the region. There was tremendous disruption, with both good and bad effects, and it is no coincidence that the Renaissance arose in its aftermath.[150] According to Norman Cantor, "[the] Black Death was the trauma that liberated the new."[151] The rational things to do in the

---

[149] Rittel and Webber, "Dilemmas," 167.

[150] Other developments such as the invention and use of the legal contract (by the Italians), which spurred trade; and a widening use of water wheels, which facilitated manufacturing, also were significant factors.

[151] Norman F. Cantor, *In the Wake of the Plague: The Black Death & the World it Made* (New York, NY: The Free Press, 2001), 202. Cantor observes that this idea is controversial, citing work by Medieval historian Dom David Knowles arguing that the Black Death had little or no impact on Europe, and that of historian David Herlihy, which argues that it was highly significant. See David Knowles, *Great Historical Enterprises and Problems in Monastic History* (London, UK: Thomas Nelson and Sons, 1962); and David Herlihy, *The Black Death and the Transformation of the West*, Samuel K. Cohn, Jr., ed. (Cambridge, MA: Harvard University Press, 1995).

14th Century were to join concentrated populations in cities with regular inter-city trade routes to move goods. These conditions provided the disruptive opportunity for the pathogen.

Pandemics are by their nature adaptive and possibly recurring. Seen in hindsight, pandemics may appear to be tame problems, seemingly clearly defined and understood. But the requirement to deal with pandemics (and other wicked problems) is not to address them merely in hindsight. Rather our well-being depends on foresight. This is, after all, how intelligence enterprises and their professionals work their issues. Do Rittel and Webber's criteria for a wicked problem provide a means of characterizing pandemics?[152]

- *Wicked problems have no definite formulation.* Over ninety years after it occurred, the 1918 influenza pandemic remains only partially understood. As Edwin Kilbourne notes, "the origin of this pandemic has always been disputed and may never be resolved."[153] Seen as it emerged, the pandemic was even less clearly understood. While germ theory was known in some places, we did not know how to apply it to the pandemic nor how to protect ourselves. Similarly, while today the causes of Avian Flu are known to be the H5N1 virus, if, when, how, and where it (or some other as yet unknown virus) mutates from an animal-to-animal (and occasionally an animal-to-human form) to a human-to-human form remains unknown. Further, the exact nature of the mutation—necessary for the formulation of a vaccine—is and remains unknown. The difficulty in determining these factors with regard to the 2009-2010 H1N1 "Swine Flu" pandemic may have led to an overestimation of the severity of the pandemic by the U.S. Centers for Disease Control.[154] While this is a problem for epidemiologists and others tasked to create vaccines, it is also a problem for the sensemakers—who have to estimate the impact of having or not having a vaccine—and the policymaker who has to consider the implications of various courses of action (with or without a vaccine).

- *Wicked problems have no clear end-point.* Specific pandemics do have an end. The disease, having run through the population,

---

152 It is germane to note that if pandemics truly are a wicked problem (as is argued here) then characterizing them as such only partially and inexactly describes them.

153 Edwin D. Kilbourne, "Influenza Pandemics of the 20th Century," *Emerging Infectious Diseases*, vol. 2, no. 1 (January 2006), 9. URL: <www.cdc.gov/eid>, accessed 28 March 2007. Cited hereafter as Kilbourne, "Influenza."

154 Carl Bialki, "Swine Flu Count Plagued by Flawed Data," *Wall Street Journal*, online edition, 23 January 2010, URL: <http://online.wsj.com/article/SB10001424052748704509704575019313343580460.html>, accessed 2 February 2010.

dissipates or abates. However, the diseases recur as the viruses evolve and mutate. In the case of the 1918 pandemic, the second wave of the virus apparently was more lethal than the first. Yet, similar, less deadly influenza occurs annually and subsequent pandemics are repeating phenomena.

- *Solutions to wicked problems are at best good or bad.* Vaccines are typically the solution to the annual influenza epidemic. In some years they are good—they are effective against the specific strains of the virus—and in some years they are not good (bad)—they are less effective against the specific strains.

- *Tests of solutions to wicked problems may not demonstrate their validity and may provoke undesired consequences.* Tests of pandemic plans and preparedness for cities and even countries provide a degree of comfort but no guarantee they will be effective against either the pandemic against which they were developed or another unexpected pandemic (such as that of the 2009-2010 H1N1 pandemic). It simply is not known nor knowable in foresight whether the measures will work, or the degree to which they will work until they are tested by the actual event. Furthermore, the proposed measures—once made—close off governments, corporations, and people from considering other options. If major adjustments are needed against the actual pandemic, it will take time to overcome the understandable resistance. Thus, testing and planning may have undesirable consequences in a real pandemic emergency.

- *Implementing solutions to wicked problems can change the problem.* One means of dealing with a pandemic in its early stages is quarantine. At the national level, this means closing international borders, preventing people as well as goods from entering or leaving the country. Doing so compounds the pandemic problem by adding economic issues. Companies relying on imported goods or on the ability to export goods may fail. Essential food items may not be available. Entire sectors of an economy may fail. The effectiveness of closing the borders depends (in part) on timing. Fewer people may die in the short-term but the longer-term economic disruptions may in fact increase mortality from other causes. The "problem" is no longer (just) the pandemic itself. Sometimes highly beneficial consequences arise out of catastrophes brought about by wicked problems. For example, as related by David Morens and Jeffrey Taubenberger, the Phillips Collection of Art in Washington, DC, owes its creation to the response of Duncan Phillips to the deaths

31

of his father and especially to the death of his brother James.[155] Phillips "as a direct consequence of the death of his brother James from influenza…dedicated his life…to establishing one of the finest public museums of modern art in the world."[156]

- *Sensemakers can never know if they have determined all the solutions to wicked problems.* Developing a collectively exhaustive list of options is difficult even when a complex issue is well understood. When the issue is not well understood, or is emerging, such lists become almost impossible to complete. In the case at hand, vaccinations, quarantines, and border closings all are suggested as means of mitigating a pandemic. However, it remains unknown what else is necessary to prevent or stem the spread of viruses and their impact on populations. While steps such as closing the borders, closing the schools, curfews, and ensuring that people wash their hands can be put in place, they can easily reach a point of unmanageability.

- *Each wicked problem is essentially unique.* While there are some commonalities between them, each influenza pandemic of the 20th Century essentially was unique. As Kilbourne notes, "Each differed from the others with respect to etiologic agents, epidemiology, and disease severity."[157]

- *Every wicked problem can be embodied in another one.* Influenza typically arises from close contact among animals and humans in agrarian settings. Thus, the influenza problem overlaps social problems that overlap economic problems and so on…

- *How wicked problems are resolved is determined by the means and methods used to make sense of them.* Means and methods of problem solving carry embedded assumptions about their appropriateness, the degree to which they are suited to the problem at hand, and of what they actually attempt to make sense. In examining a pandemic, vaccination strategies lead to certain results, whereas border closures lead to other ones.

- *Sensemakers have no right to be wrong.* Nor, in the case of pandemics or other global threats, do their policymaking consumers. When the stakes are the lives of many people, sensemakers

---

[155] David M. Morens and Jeffery K. Taubenberger, "Influenza and the Origins of The Phillips Collection, Washington, DC," *Emerging Infectious Diseases*, vol. 2, no. 1 (January 2006), 79. URL: <www.cdc.gov/eid>, accessed 28 March 2007. Cited hereafter as Morens and Taubenberger, "The Phillips Collection."

[156] Morens and Taubenberger, "The Phillips Collection," 79.

[157] Kilbourne, "Influenza," 9.

and policymakers who miscalculate or underestimate or are otherwise wrong about a pandemic and its impact on their countries or region can expect vilification at best. A fear of such vilification from the public and the media might contribute to the situation whereby pandemic-tracking organizations such as the U.N. World Health Organization or the CDC *overestimate* the severity and threat posed by a pandemic such as the 2009-2010 Swine Flu pandemic.[158] For intelligence professionals this phenomenon is not unknown. Common wisdom among intelligence sensemakers is that it is far better to warn and be mistaken (and nothing happens) than to not warn and be mistaken (something happens).

Complexity

Rittel and Webber's notions of wicked problems can also be characterized through the lens of complexity theory. As considered by Jonathan Rosenhead, "systems of interest to complexity theory, under certain conditions, perform in regular, predictable ways; under other conditions, they exhibit behaviour [sic] in which regularity and predictability is lost."[159] This is exceptionally true of intelligence. Certain kinds of issues, including the interpretable indications of a build-up to armed conflict, can be extremely predictable. For example, if one observes a mass of troops approaching a national border and knows that the means by which these troops were trained includes a doctrine of "mass and attack," then one might legitimately adduce that an attack is likely and imminent. One could even use the past as a means of prognosticating the future with some degree of legitimate confidence.

However, in other situations, there may be a number of unknowable, unpredictable, and unanticipatable outcomes. Thus, reliable prognostication is simply not possible.[160] For instance, if a coalition of nations removes an oligarch in another nation from power, the specific outcomes of that action cannot be known in foresight. While alternative outcomes can be modeled

---

[158] Whether overestimation occurred was argued in the European Parliamentary Assembly, although the WHO denied that it occurred. See "WHO Rejects Accusations It Mishandled H1N1, Updates Worldwide Stats," Kaiser Family Foundation, URL: <http://globalhealth.kff.org/Daily-Reports/2010/January/25/GH-012510-Swine-Flu.aspx?utm_source=feedburner&utm_medium=feed&utm_campaign=Feed%3A+kff%2Fkdghpr+%28Kaiser+Daily+Global+Health+Policy+Report%29>, accessed 2 February 2010.

[159] Jonathan Rosenhead, "Complexity Theory and Management Practice," URL: <http://www.human-nature.com/science-as-culture/rosenhead.html>, accessed 23 December 2008. Cited hereafter as Rosenhead, "Complexity Theory." See also, Jonathan Rosenhead, ed., *Rational Analysis for a Problematic World: Problem Structuring Methods for Complexity, Uncertainty and Conflict*, 2nd Edition (Hoboken, NJ: John Wiley & Sons, 2001).

[160] More on this below.

and simulated, they remain valuable only as discussion points: there is no guarantee in advance that they have captured the reality that will occur. Modeling and simulation are feasible because complexity science shows that the "indeterminate meanderings of these systems, plotted over time, show there is pattern in the movements…the pattern stays within a pattern, a family of trajectories."[161] Unfortunately, because intelligence must address the "real" world, rather than its modeled or simulated semblance, events often are unique and therefore their patterns also are unique.[162] Thus, there exists an inability to guarantee a future reality; even probabilities may be suspect.

Analysis as here defined is insufficient to address complexity. Disaggregation simply does not reveal future alternatives. That this is so becomes obvious if one finds that it is the emergence of unique and novel behaviors arising from different and minutely differing initial conditions that characterize many 21st Century intelligence issues. In these circumstances, the whole of an issue is *greater* than its parts. But, in analysis, the issue is by definition and practice the *sum* of its parts.

Given these complex issues, the concept of "analysis" is simply insufficient for sensemaking. Instead, greater conceptual accuracy and precision of terminology is required. To achieve the needed accuracy and precision requires more than semantic invention. It also demands that underlying concepts, known as assumptions or premises, be identified and accounted for. Therefore, in developing the case for considering new paradigms for intelligence, certain terms require explicit (re)definition.

Implications of Complexity

Viewed from a larger context, complexity stymies the entire "standard model" of intelligence creation. With regard to Kent's concept of knowledge, or how intelligence is created, complexity—as viewed from the framework of wicked problems—confounds the consideration and mitigation of such problems. Kent's model of predictive and specific warning seems more miss than hit. Complexity further confounds the collaborative processes contained

---

[161] Rosenhead, "Complexity Theory."

[162] It should be noted that some intelligence-associated phenomena do exhibit general patterns that may be indicative. Observations of military preparations are an example, although caution is necessary when one extrapolates what those observations mean. For example, while one may make certain conclusions about a massing of troops on a border in light of who trained those troops (as was the case with Iraqi troops massed along the Kuwaiti border in 1990), what the indicator means may be subject to error. This latter point is illustrated by CIA Director John McCone's conclusion that SA-2 deployments in Cuba indicated strategic nuclear missiles were also being deployed. McCone's conclusion—although it turned out to be correct in the Cuban case—ignored the deployment of SA-2s without (presumably) the accompanying strategic nuclear-armed missiles to Egypt and Syria that was taking place at the same time. See Moore, *Critical Thinking*.

within Kent's notions of Activity and Organization, by which intelligence professionals are tasked to interact. How does the intelligence professional know in advance whose imagination will be most helpful in making sense of the problem at hand in time to prevent a catastrophe or even imagine one? For example, viewed in hindsight, the Drug Enforcement Administration (DEA) might have had a key piece of knowledge that would have been useful in the early apprehension of the December 2009 "underwear bomber." Certain illegal drugs—marijuana is one—are often "stashed" in a traveler's underwear.[163] In retrospect, one question becomes, "what else could be carried in this fashion?"

However, asking such questions in retrospect is irrelevant. The question should be asked in foresight—but how does an intelligence professional or even an airline security official know to ask such questions in advance? Here again, complexity confounds the necessary collaboration. An intelligence professional might ask, "Which agency or agencies can help me make sense of this issue?" With 17 intelligence agency partners, there are up to 131,071 possible collaborative combinations—any of which (in various combinations) might be valuable.[164] This number presumes the searching intelligence professional knows which person—and there is only one person—in each agency who can help. The number grows at least exponentially when more than one individual at each agency might be helpful. While this situation presumes that the intelligence professional does not know who that person is in advance, in fact the professional does have a likely list of contacts, reducing the number of possible combinations dramatically. Still, this collaboration exercise may be a wicked problem. There may be sets of better collaborative combinations. The key imagination—as characterized by the DEA in the example above—may not be part of the collaborating group. One challenge is to determine a best collaborative combination prior to the event and develop believeable scenarios of what might transpire.[165] Finally, it

---

163 The author presumes the DEA is aware of this practice and would think of it if asked "how might one transport something without its being detected." The author's youthful observations while growing up in the American Southwest as well as discussions with state and local law enforcement officials engaged in intelligence-led policing training suggest the practice is fairly widespread.

164 The formula for such calculations is 2 raised to the power of the number of variables, in this case intelligence agencies, or 17, minus 1 ($2^n - 1$). See David T. Moore and William N. Reynolds, "So Many Ways To Lie: The Complexity of Denial and Deception," *Defense Intelligence Journal*, vol. 15, no. 2 (Fall 2006): 95-116. While the author and William Reynolds discuss complexity calculations in the context of denial and deception, the same notions apply to collaboration.

165 Peter Schwartz and Doug Randall, "Ahead of the Curve: Anticipating Strategic Surprise," in Francis Fukuyama, ed., *Blindside: How to Anticipate Forcing Events and Wild Cards in Global Politics* (Washington, DC: Brookings Institution Press, 2007), 94. Cited hereafter as Schwartz and Randall, "Ahead of the Curve."

must lead to an action. As Peter Schwartz and Doug Randall write, "Achieving believability and action requires a depth of insight and understanding that is rare."[166]

Given the challenges of both tame and wicked 21st Century intelligence problems and their inherent complexity, what are intelligence professionals to do? One avenue open to them, and presented below, is the development and validation of methods of reasoning about key evidence.

---

[166] Schwartz and Randall, "Ahead of the Curve," 94.

CHAPTER 3
From Normal to Revolutionary Intelligence

Evidence-Based Intelligence Creation

Intelligence sensemakers use more than context-less data and information. They employ assemblages of *evidence*—at a minimum, collections of data and information determined through marshaling to be relevant to the issue under consideration—in other words, contextualized to specific issues. Evidence reveals alternative explanations through pattern-primed, induced inferences about what is going to happen or what has happened already in the past.[167] While the inferences are typically uncertain, they do justify beliefs about phenomena. Justifying beliefs (or theories or hypotheses) presents a case for their accuracy but does not guarantee ground (or any other) "truth." Rather, as Peter Kosso notes, justifying beliefs is "about meeting the standards of evidence and reason [to] indicate [the] likelihood of accuracy."[168] Sensemakers go further and seek to demonstrate that the knowledge of tendencies they establish provides for "a correlation between being more justified and being true."[169] For example, during a recent exercise in which the author was a participant, one team reached an inferential conclusion about a likely explanation of the phenomenon being examined. The participants were presented with a set of previously determined alternative conclusions and a set of supposedly relevant evidence and asked to assess which (if any) conclusions were justified and true. After reaching an initial position they then were required to consider an alternative conclusion that presumed their original conclusion was false. In so doing they were surprised to find that while the original

---

[167] Reasoning about past events remains easier than reasoning about the future. In the former case, the evidence may be contradictory, deceptive, and subject to more than one interpretation. However, it tends to be more complete and marshaled (or at least discoverable). When looking to the future the important consideration is that much of the evidence does not yet exist. The events described by precursor information may not have occurred. Additionally, the tests of the information that transform it into evidence necessarily remain incomplete.

[168] Peter Kosso, "Introduction: The Epistemology of Archaeology," in Garrett G. Fagan, *Archaeological Fantasies: How Pseudoarchaeology Misrepresents the Past and Misleads the Public* (London, UK: Routledge, 2006), 4. Cited hereafter as Kosso, "Epistemology."

[169] Kosso, "Epistemology," 4.

conclusion initially appeared more accurate, the latter one was in fact "true." They reached their incorrect initial conclusion in part because they made wishful assumptions about the evidence they used to justify that conclusion. As Kosso observes, "more justification is better, since it raises the likelihood of accuracy. But it is certainly possible for a well-justified belief to be false."[170] This realization on the part of the participants subsequently led to a more critical assessment of the evidence, much of which was found to be false. They discovered that "justification comes in degrees, but truth does not."[171]

It is arguable whether greater evidentiary justification demonstrates the likelihood of a strongly accurate correlation with truth. As Kosso makes clear, even with abundant justification, there is no certainty of truth. Therefore, according to Kosso, it "is the task of the systematic disciplines…to refine carefully the content of justification, the evidence and the network of theoretical beliefs, to bring justification into ever closer correlation with truth."[172] If intelligence is to "speak truth to power" it must first ensure its words are well and critically justified.[173]

As figure 1 illustrates, intelligence sensemaking is conducted in service of a number of goals, including describing states of affairs, explaining phenomena, interpreting events and actions, and estimating the likelihood and impact of a foe's future actions As intelligence professionals move from describing events, explaining patterns of behavior, and grasping underlying factors and intentions, ever more justification of beliefs about the phenomena under scrutiny is required. Yet, as intelligence professionals attempt to apply greater scrutiny in this sequence, their capability to do so decreases as they face greater ambiguity.

Additionally, we may expect that sensemakers will more often be wrong in offering *estimations* about phenomena than when they are merely describing them. This is in part because of an interesting reality characterized by Taleb: To predict the future we must already know the future.[174] What Taleb means is that one has to already have visualized what the future will be in order to estimate it. Kosso, in writing about epistemic science, elaborates:

---

[170] Kosso, "Epistemology," 4.

[171] Kosso, "Epistemology," 4.

[172] Kosso, "Epistemology," 5.

[173] The aphorism "speak truth to power" is variously (and wishfully) ascribed to Sherman Kent. Research by the author and discussions with former intelligence officers who knew and worked with Kent fails to reveal this to be the case. Intelligence strives to ensure its findings are factual and—to the best knowledge of its creators—"true."

[174] Nassim Nicholas Taleb, *The Black Swan: The Impact of the Highly Improbable* (New York, NY: Random House, 2007), 173. Cited hereafter as Taleb, *The Black Swan*.

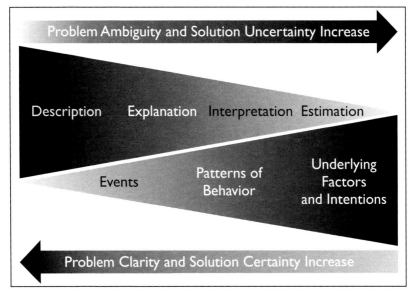

Figure 1. Types of Intelligence and the Phenomena They Characterize.
Complexity, ambiguity, and uncertainty increase as one moves from Descriptive to Estimative Intelligence.
Source: Author.

> [The] knowledge claims are more ambitious in that they stray further from what is immediately observed…The theoretical descriptions are based on observation and evidence…But it is important to note that the observations themselves are based on theory. Scientific evidence, after all, is neither haphazard nor uninterpreted, and some prior conceptual understanding of nature will inform decisions about what to observe, which observations are credible, what the observation means, and how what is observed is causally (and hence intentionally) linked to what is not observed. Theory is necessary to turn mindless sensations into meaningful evidence.[175]

While Kosso writes in the language of science, his argument applies as well to Manjoo's "post-fact" world of which intelligence often attempts to make sense.[176] Intelligence is created ultimately from human, sensory-mediated observations of phenomena. Further, intelligence evidence, while

---

175 Kosso, "Epistemology," 5.

176 By "post-fact" Manjoo refers to a tendency to ignore evidence in favor of predetermined conclusions. Common examples include the arguments of pseudosciences such as astrology or intelligent design. A more serious example is manifested by parents who refuse—due to pseudoscientific (and disproved) conclusions—to vaccinate their children. Within the IC this tendency manifests itself when intelligence professionals let predetermined conclusions drive their intelligence creation efforts rather than let evidence go in search of hypotheses.

it may appear to remain "haphazard," is the result of systematic foraging, gathering and interpretation. The past tells intelligence practitioners what to look for in the future. This poses dangers when those indicators are no longer (if indeed they ever were) valid. As Baruch Fischoff suggests, "searching for wisdom in historic events requires an act of faith—a belief in the existence of recurrent patterns waiting to be discovered."[177] Yet, while general patterns may exist, "the past never repeats itself in detail."[178] In other words one might detect indicators suggesting that an upcoming event similar to one in the past is possible, likely, or even reasonable. On this basis, one could, for example, have anticipated that sooner or later foreign terrorists would again attack the United States by targeting some high-value building or event, such as the World Trade Center.[179] This is a far cry from predicting that Al Qaeda terrorists would fly airplanes into the World Trade Center and the Pentagon on the morning of 11 September 2001.[180] Finally, intelligence practitioners making sense of issues rely heavily on theories, which if unacknowledged, essentially are unexamined assumptions.

If using the past to gain wisdom about what the future holds is not feasible, what about studying the past to avoid folly? Tversky and Kahneman's work on *availability* leads one to suspect (as Fischoff also notes) that focusing on misfortunes "disproportionately enhance[s] their perceived frequency."[181] Another challenge to considering the past as a teacher of what not to do is that one may not properly understand the problem. While understanding may be possible (or even easier) when dealing with Tame Problems—as has been discussed above, when considering Wicked Problems, such understanding is elusive at best and is dependent on the methods chosen to deal with the problem; in other words, woefully incomplete.

---

[177] Baruch Fischoff, "For Those Condemned to Study the Past: Heuristics and Biases in Hindsight," in Daniel Kahneman, Paul Slovic, and Amos Tversky, *Judgment Under Uncertainty: Heuristics and Biases* (Cambridge, UK: Cambridge University Press, 1982), 338. Cited hereafter as Fischoff, "Condemned to Study the Past."

[178] Fischoff, "Condemned to Study the Past," 336.

[179] On 26 February 1993, a truck bomb placed by Islamic terrorists exploded below one of the buildings of the World Trade Center and was apparently intended to topple both towers. For a discussion of the security implications, see Laurie Mylroie, "Who is Ramzi Yousef? And Why It Matters," *The National Interest*, 22 December 1995. Available at URL: <http://www.fas.org/irp/world/iraq/956-tni.htm>, accessed 12 October 2010.

[180] Such an attack had been *anticipated*. As Schwartz and Randall comment, "many people did anticipate the terrorist attacks of September 11 [2001]...Yet most Americans, as well as officials in both the Clinton and Bush administrations, focused their attention elsewhere while the inevitable grew imminent." Schwartz and Randall, "Ahead of the Curve," 94.

[181] Fischoff, "Condemned to Study the Past," 339. See also Amos Tversky and Daniel Kahneman, "Availability: A Heuristic for Judging Frequency and Probability," *Cognitive Psychology*, vol. 5 (1973): 207-232.

With the intention to improve evidence-based intelligence creation, recent legislation "reforming" intelligence goes so far as to require that "alternative analysis" be conducted.[182] The IC, at least through its schools, interprets this to mean that *multiple* hypotheses be considered. The relevant act mentions "red teaming: a means by which another group of intelligence professionals consider alternative explanations for an issue being scrutinized."[183] The legislation leaves unexamined the question of whether the criteria for sensemaking will be met in examining tame problems and especially wicked problems arising from consideration of adversarial intentions.

If, for example, one estimates that a particular country whose policies one's own government generally opposes will develop both a long-range missile capability and a nuclear weapons capability and then marries the two together, one has to have already imagined, in the context of the target country's political and technological environment, what a long-range missile capability is, what a nuclear weapon is, what a weapon of mass destruction is, and a strong sense of the will to combine these threat elements. Policymakers may challenge the target country's actions, making their leaders more adversarial. Thus, at a minimum, intelligence and policy create the future—or a version of it. Done poorly, this can lead to unintended and dangerous implications.

Emerging from this mélange are hypotheses that are ripe for disconfirmation, although a tendency to compound uncertainties on the part of intelligence professionals and their consumers may serve to prevent this from occurring. If in fact, in the example above the assessment is wrong, and the nuclear capability assessed to be for weapons really is intended to provide an alternative to the nation's dependency on a dwindling supply of increasingly more expensive petroleum as a source of energy, and the missiles support a nascent space program designed to orbit telecommunications satellites that can fulfill the country's needs and those of their neighbors as well as generate income for the government, then sensemakers and policymakers on all sides will have extrapolated inappropriate patterns arising from a misunderstood present. In a tense bilateral or even multilateral environment, rhetoric and actions can precipitate events so as to create a future consistent with those pattern-derived conclusions, driving the target country to produce the weapons. Each side then blames the other nation's government for having "caused" the crisis.

---

182 U.S. Congress, IRTPA, 2004, 33.
183 U.S. Congress, IRTPA, 2004, 33.

There are a number reasons why such faulty causal networks occur. Honest evidentiary considerations demand a degree of agnosticism about the theories being justified. Evidence-based knowledge, as Peter Kosso's epigraph at the beginning of this book acknowledges, is not absolute; justified evidence *changes* theories and not the other way around. In other words, when interpretations of the evidence lead to coherent alternative inferential conclusions, then the existing or accepted theories require changing. What must not happen is to reinterpret the evidence to support the prevailing pre-existing theory, as is the case in the above example.

However, this is exactly what happens all too often. People are often unwilling to abandon their cherished positions. This occurs in part because they are not dispassionate as they reason about evidence. In other words, positions are influenced by various worldviews or cognitive approaches, particularly selective perception and selective exposure. These combine to steer how people recognize issues, the phenomena that comprise them, and how they go about making sense of them.[184] These influences or theoretical frameworks shape the patterns people use to interpret new phenomena. The benefit is that these frameworks make people smart and do so quickly.[185] However, this benefit depends on the existence of a repetition of observed phenomena. In order to work successfully, an intuitive framework for understanding requires at least a similar situation, a condition that may not occur with intelligence phenomena.

In an information-rich environment brought about by technical collection, intelligence professionals can select inappropriate patterns to use in making sense of new phenomena. In intelligence work, if such patterns conspire to affect the search for and the selection of the evidence sensemakers use, and that they and their consumers then accept, selective perception and selective exposure set the stage for intelligence error and failure.

Evidence always requires a context, and as the missile example illustrates, there may be more than one explanatory context that makes sense. In intelligence, "evidence is [particularly] rarely self-sufficient in information or credibility."[186] Additionally, the dispassionate nature of evidence itself, when

---

[184] For more information on frameworks or "heuristics" that people employ to cope with judgments about which they are uncertain, the reader should consult the works of Daniel Kahneman, Paul Slovic, and Amos Tversky, *Judgment Under Uncertainty: Heuristics and Biases* (Cambridge, UK: Cambridge University Press, 1982) and Thomas Gilovich, Dale Griffin, Daniel Kahneman, *Heuristics and Biases* (Cambridge, UK: Cambridge University Press, 2002).

[185] This is the premise of Gerd Gigerenzer's book *Simple Heuristics That Make Us Smart* (Oxford, UK: Oxford University Press, 1999). These patterns of behavior evolved as survival mechanisms and by and large they have sufficed—the human species survives to the present.

[186] Kosso, "Epistemology," 8.

viewed outside its political and social context, contributes to a failure to adequately explore issues. In other words, unless the correct context is known, evidence—if its constituent information can even meet that threshold—is subject to many different interpretations. Without context the person assessing the evidence has no way of knowing which interpretation is correct. Multiple contexts further confound the situation, for different contexts often lead to alternative conclusions as was illustrated in the missile development scenario just described. Finally, as Hampson's essay reveals, the political context of the policymaker may skew the actual context conveyed by intelligence.[187]

What is occurring in the contextual consideration of evidence is a process of epistemological justification and as Kosso notes the "key concern is to distinguish knowledge, on the one hand, from mere belief, opinion, dogma, and wishful thinking, on the other."[188] In relation to intelligence, knowledge depends on contextual justification of evidence and, as noted earlier, the "business of epistemology is to show that there is a correlation between being more justified and being more likely to be true."[189]

Despite their inherently greater inaccuracy, predictions seem to garner more interest from consumers than do explanations. In biology, predictive hypotheses require accommodation to valid background information in order to be useful.[190] Is this true in intelligence? As illustrated by two National intelligence Estimates (NIE's) dealing with weapons of mass destruction (WMD) that were subsequently found to be wrong, *incorrect* predictions certainly garner considerable attention. In these cases, the incorrect predictions perhaps resulted in part from a lack of time available for their preparation. In the first case, the estimate on whether or not the Soviets would place strategic nuclear missiles in Cuba in 1962 was written in a week.[191] In the case of Saddam Hussein's WMD programs, the preparation time was three weeks. Such short time frames for preparation would seem to prevent new information and data

---

187 Hampson, "The Divided Decision-Maker."

188 Kosso, "Epistemology," 4.

189 Kosso, "Epistemology," 4.

190 Kathrin Stanger Hall and others, "Accommodation or Prediction?" Letter in response to Peter Lipton, "Testing Hypotheses: Prediction and Prejudice," *Science*, vol. 308, no. 5727 (3 June 2005), 1409.

191 Admittedly this is hindsight analysis. Sherman Kent, writing in 1964, asserts that in retrospect the authors *did* have sufficient time to assess the evidence. See Kent, "A Crucial Estimate Relived," *Studies in Intelligence*, vol. 8, no. 2 (Spring 1964):1-18. Originally classified SECRET, it was declassified and reprinted in *Studies in Intelligence*, vol. 36, no. 5 (1992): 111-119. Kent carefully considers the reasons why the estimate was wrong and why it was not revised. Among the reasons cited was the presumption that past precedents of Soviet foreign policy would continue into the future. Thus, no offensive missiles would be deployed in Cuba.

from being collected and made relevant to the issues (i.e. marshaled as evidence), and prevented alternative perspectives from being fully explored.[192]

A lack of time typifies one context of intelligence sensemaking. This context in which intelligence professionals work and the constraints imposed upon them facilitate their successes but also their failures. As the above examples demonstrate, intelligence sensemakers are often under pressure to consider massive amounts of data and information in a short time. These same professionals must marshal the data and information into evidence even as they attempt to understand, and then explain, the associated complex issues to policymaking and decisionmaking consumers.

Finally, the current and past practice of intelligence in the U.S. is consistent with a focus on prediction rather than explanation as its ultimate purpose. This has not been without critique. Willmoore Kendall, in reviewing Kent's book in 1949 for *World Politics*, criticized Kent's "compulsive preoccupation with *prediction.*"[193] Given the experience of Kent and others during World War II, it is not surprising that the desire to prevent another Pearl Harbor dominated their practices; such a desire naturally would have led to an activity organized around the creation of surprise-preventing predictive knowledge.[194]

Considering the Normal

The process described in the preceding section can be thought of as "normal intelligence." As conceived by Thomas Kuhn, "normal" refers to "the relatively routine work…within a paradigm, slowly accumulating detail in accord with established broad theory, not actually challenging or attempting to test the underlying assumptions of that theory."[195] We can thus see that "normal intelligence" is an activity of expanding knowledge in which most intelligence professionals engage and which incrementally increases knowledge about targeted phenomena.[196] The professionals work within a model

---

192 The Senate report on the Iraqi WMD noted that alternatives were not explored although this is not the case with the aluminum tubes alternately argued to be for centrifuges and rockets. In this case a groupthink framework seems to have shaped the results. See United States Senate, *Report on the U.S. Intelligence Community's Prewar Intelligence Assessments on Iraq*, Select Senate Committee on Intelligence, 108th Congress, 7 July 2004, 18, 21.

193 Kendall, "Function of Intelligence," 549.

194 Given a "worst-case" perspective about WMD such a perspective perhaps also partially explains points of view about Iraq's alleged WMD programs leading up to the 2002 estimate. It does not, however, explain the apparent 9/11 "failures of imagination."

195 Wikipedia, entry under "Normal Science," URL: <http://en.wikipedia.org/wiki/Normal_science>, accessed 26 September 2007.

196 Thomas Kuhn, *The Structure of Scientific Revolutions* (Chicago, IL: University of Chicago Press, 1962), 10-42, Cited hereafter as Kuhn, *Structure*.

or "paradigm" of reality forged during the Second World War and reinforced throughout the ensuing Cold War. The perceived and recalled successes of the past contribute to the repeat use of unvalidated tradecraft. The paradigm presumes state-level adversaries—eventually with mutually destructive capabilities.

As used in this context, "normal intelligence" is to "intelligence" as Thomas Kuhn's "normal science" is to "science." In both domains newly created knowledge incrementally adds to an increasingly established paradigm; new knowledge does not easily redefine the paradigm. Even the anomalous and truly unknown are only considered in terms of what is known. Normal science or normal intelligence does not seek to revise significantly the paradigm by which new phenomena are known and understood. This may be seen in the way new intelligence personnel adopt existing job accounts. A common practice involves their reviewing previous reporting on the account, with a tendency for new reporting to stay within the conceptual boundaries of what has gone before. Knowledge increases only incrementally.

Intelligence practitioners working within the paradigm of normal intelligence attempt to describe, explain, or predict phenomena *coherently*. In this case, the term refers to another concept developed by Kuhn: A common framework for understanding the phenomena is sought or presumed. New knowledge is understood in the context of the dominant paradigm. For example, normal intelligence of the latter half of the 20th century understood events in relation to the missions and goals of U.S. adversaries, principally the former Soviet Union and to a lesser degree, China. World affairs were understood in the context of the state-based adversaries' hegemonic competition with the United States.

Normal paradigms prevail until previously unnoticed and unnoticeable discrepancies create sufficient inconsistencies in explaining and understanding phenomena so as to cause errors that cannot be ignored. For Kuhn this means a necessary change in scientists' perception of the realities as characterized by science. Kuhn illustrates this with an example of the

> transition from Ptolemaic to Copernican astronomy. Before it occurred, the sun and moon were planets, the earth was not. After it, the earth was a planet, like Mars and Jupiter; the sun was a star, and the moon was a new sort of body, a satellite. Changes of that sort were not simply corrections of individual mistakes embedded in the Ptolemaic system...they involved not only changes in laws of nature but also changes in the criteria by which some terms in those laws are attached to nature. These criteria, furthermore,

were in part dependent upon the theory with which they were introduced.[197]

In the physical sciences, the phenomena themselves do not change (although newly noticed phenomena could make it appear so). In the cultural environment of human interaction, the *new perceptions* of reality can be enough to force a reconsideration of the *old*. In social scientific terms, a new paradigm not only explains the new, it does *better* at explaining the old. Further, even the language previously employed to describe a phenomenon is inadequate because—as Kuhn notes—"scientific development cannot be quite cumulative. One cannot get from the old to the new simply by an addition to what was already known."[198]

Failures to consider discrepancies lead prospective intelligence sensemakers to retain an invalid understanding of phenomena even as the phenomena themselves change. This larger discrepancy leads either to intelligence error or intelligence failure. Once again, within the Kentian paradigm, intelligence errors derive from "factual inaccuracies in analysis resulting from poor or missing data."[199] Conversely, as has been noted, intelligence failures refer to "systemic organizational surprise resulting from incorrect, missing, discarded, or inadequate hypotheses."[200] Within the former concept there is a presumption that if more data are available or better understood, errors can be prevented. In the latter, intelligence practitioners or their policymaking customers have misunderstood the issue and its context.

The existence of particular intelligence errors *does not* necessarily indicate a paradigm has changed. However, repeated intelligence errors *do*. As is the case with science, small errors in adequately characterizing phenomena lead to the emergence of "corrective constants." The sensemaker may have made a perceptual or interpretive error. However, left unchecked, errors eventually combine to cause systemic failures. Intelligence practitioners and policymakers repeatedly may come to incorrect conclusions from faulty sensemaking, leading to policy failures, defined by Rob Johnston as "systemic organizational surprise" resulting from a mixture of practitioners' lapses and policymakers' ignoring proffered intelligence.[201] The more

---

197 Thomas S. Kuhn, *Thomas S. Kuhn, The Road Since Structure: Philosophical Essays, 1970-1993, with an Autobiographical Interview*, James Conant and John Haugeland, eds. (Chicago, IL: University of Chicago Press, 2000), 15. Cited hereafter as Kuhn, *The Road Since Structure*.

198 Kuhn, *The Road Since Structure*, 15.

199 Johnston, *Analytic Culture*, 6.

200 Johnston, *Analytic Culture*, 6.

201 Johnston, *Analytic Culture*, 6.

germane interpretation offered in the present work is that intelligence failure stems from inadequacies on the part of both policymaker and intelligence professionals to recognize a fundamental, revolutionary shift of the intelligence context.

The state-as-adversary paradigm for intelligence creation is obsolete. Two decades now separate the interpretable intelligence context from that of the Cold War: the adversaries and issues are now strikingly different.[202] The power of the Soviet Union waned dramatically after 1990 as that of China increased. But even more central to the intelligence context, novel phenomena also appeared that were non-state based: emerging non-state actors posed new challenges by threatening traditional state structures. Emerging phenomena such as disease and climate change imposed new complexities. The anomalies these new phenomena have created illustrate how and why normal intelligence is no longer adequate: it could no longer characterize these phenomena within the threat and opportunity framework of strategic intelligence. The "normal" means by which error is explained remain inadequate. As documented in the various Congressional and independent commission reports, intelligence no longer adequately describes, explains, or predicts with respect to the phenomena its consumers need to understand. Thus, intelligence change is necessary—revolutionary change.

Paradigm Shift

Revolutions in science, politics, and military affairs occur because crises reveal the insufficiency of the reigning paradigm. As Kuhn notes, an existing paradigm can "cease to function adequately in the exploration of an aspect of nature to which the paradigm itself [has] previously led the way."[203] Kuhn argues further that science does not evolve smoothly. Rather, periodic revolutions change how phenomena are perceived and understood.[204] Crises are a precursor of such paradigm shifts. Analogously to Kuhn's notions, the (officially) serious failures of the Intelligence Community to predict the attacks of 11 September 2001 and the state of Saddam Hussein's programs of

---

[202] It should be noted that the traditional adversaries of the Cold War remain threats. While an argument can be made that other issues overshadow the dangers they pose, it can also be claimed that emerging threats simply compound the traditional ones. 21st Century dangers are complex; one of their dangers *is* their complexity.

[203] Kuhn, *Structure*, 92.

[204] Kuhn, *Structure*, 92-110.

WMD (both in 1990 and prior to the 2003 invasion of Iraq) are examples of systemic reframing crises.[205]

If we review these failures through their official re-examinations, certain phrases recur: mind-sets, politicization, and faulty analysis. Across the same period, there are repeated attempts to impose methods of "[social] scientific study...to analysis of complex ongoing situations and estimates of likely future events."[206] What is lacking is any sort of a systematic approach across the Intelligence Community. As long-time practitioner and observer Jack Davis noted over a decade ago, no corporate standards for how intelligence is created, including the methods employed, exist.[207] Although sound practice does not ensure that intelligence assessments will be correct, its absence, by definition, contributes to flawed conclusions. Contributing to this scenario is the fact, as Aris Pappas and James Simon observe, that "[potential] opponents [are] often...driven by emotional agendas that [make] them unpredictable."[208] While effective practice might not drive the production of sound estimates from ambiguous evidence, it would routinely alert practitioners to the prospect that these same opponents' actions are unpredictable or at best, that they are calculable within a range of behaviors.

In short, U.S. intelligence professionals operate in an environment similar to an unfolding Kuhnian revolution: the epistemology of normal intelligence is insufficient and new knowledge is needed. The recent failures highlight the necessity for change, as does the graying of the intelligence sensemaking workforce—new people faced with new and emerging issues should be comfortable with finding new ways to systematize their work. The changed contexts and data, once they confront practitioners with problems that are

---

[205] The argument that these were crises in reframing rests on the observation that intelligence practitioners were as shocked at the inaccuracy of their estimates as were the surprised policymakers. In the case of the 2000 estimates of Iraqi WMD, the CIA *underestimated* the state of Iraqi WMD, leading to a claim that the apparent overestimation in the 2002 estimate was compensation for that earlier error. While admitting to the underestimation, a CIA press release adds, "in no case were any of the judgments [in the 2002 estimate] 'hyped' to compensate for earlier underestimates." See, Central Intelligence Agency, "Iraq's WMD Programs: Culling Hard Facts from Soft Myths," *Press Release*, 28 November 2003, URL: <https://www.cia.gov/news-information/press-releases-statements/press-release-archive-2003/pr11282003.html>, accessed 9 December 2009.

[206] Jack Davis, "Introduction—Improving Intelligence Analysis at CIA: Dick Heuer's Contribution to Intelligence Analysis," in Richards J. Heuer, Jr., *Psychology of Intelligence Analysis* (Washington, DC: Center for the Study of Intelligence, 1999), xv. Cited hereafter as Davis, "Improving Intelligence."

[207] Davis, "Improving Intelligence," xxv.

[208] Aris A. Pappas and James M. Simon, Jr., "The Intelligence Community: 2001-2015: Daunting Challenges, Hard Decisions," *Studies in Intelligence*, vol. 46, no. 1 (2002), URL: <http://www.cia.gov/csi/studies/vol46no1/article05.html>, last accessed 10 January 2006.

unintelligible in normal intelligence, will reflect the idea that a Kuhnian-style revolution in intelligence is underway.

However, a caveat is necessary. Not all "old school" intelligence practices are without continuing value. Several significant state-level adversaries remain as threats to the security of the American nation although they too are challenged by the new non-state actors and issues that populate the paradigm of the new intelligence—something that compounds any estimate of how they are likely to engage the United States. Further, in many circumstances and in dealing with certain issues, the tacit expertise of highly experienced intelligence professionals is appropriately tapped for "recognition-primed" sensemaking.[209] These "old hands" possess both current knowledge and a highly evolved skill set. Years of innovative and critical thinking mean they are skilled in looking at issues from a variety of perspectives and have the wisdom of deep context. It is no accident, therefore that the contributors to Roger George and James Bruce's 2008 book, *Analyzing Intelligence*, are very senior intelligence practitioners.[210] The challenges involve *knowing* when such expertise is valuable and needed in the first place, and encouraging the intelligence enterprise to *develop* and *retain* the cognitive and organizational flexibility that such thinking requires.

Indeed, a part of successful and revolutionized intelligence work involves gleaning new meanings from old patterns that have remained hidden to those who have stopped short of sensemaking. One challenge is that the "fresh" eyes lack the knowledge of potentially relevant patterns while the "old" eyes cannot see things as new. Each lacks the other's strength. Experience acquired by newer professionals who engage in the practice of traditional "analysis" jaundices their once-fresh viewpoints even as they start to acquire the relevant and necessary experience.

One solution may be to adopt a model of core competencies broken out according to task analyses of existing intelligence missions and functions. Such a model identifies what is needed and has been at least partially implemented in the IC's Analytic Resource Catalog developed during the tenure of former Director of Central Intelligence George Tenet.[211] Rewarding the successful use of some of the most important competencies may also encourage

---

209 Robert Hoffman, conversation with the author, 4 October 2007.

210 Roger Z. George and James B. Bruce, *Analyzing Intelligence: Origins, Obstacles, and Innovations* (Washington, DC: Georgetown University Press, 2008).

211 Mark M. Lowenthal, "Foreword," in Moore, *Critical Thinking*, xi.

their retention in the catalog. Among these are curiosity, perseverance, and pattern recognition.[212]

A necessary first step in a revolution in intelligence work is looking in depth at what it is intelligence professionals do, must do, and how they do it. Simply put, intelligence practitioners create knowledge to support their customers. As used here, intelligence practitioners are presumed to be contributors to government plans and policies at a variety of levels where they have the opportunity to share broad strategic perspectives with national leaders as well as ensure that deployed warfighters have at hand the fruits of technical collection and marshaling of tactical data.

Finally, it should be noted that intelligence *Knowledge* is only one component of a strategic, operational, and tactical intelligence triumvirate.[213] *Activity* and *Organization* are the other two. It is the author's belief that Activity and Organization also are in need of new paradigms. However, such a discussion hinges on what intelligence Knowledge is and how it is created—in short, the sensemaking involved. Insofar as Activity describes how the precursors of intelligence are hunted, gathered, made sense of, and transformed into knowledge, it is considered here. However, the uses of intelligence (also an activity) and how intelligence professionals are grouped, led, and managed to act and create knowledge—the realm of Organization—lie beyond the scope of this book.

---

[212] For more on core competencies for successful intelligence work see David T. Moore and Lisa Krizan, "Core Competencies for Intelligence Analysis at the National Security Agency," in Russell G. Swenson, ed., *Bringing Intelligence About: Practitioners Reflect on Best Practices* (Washington, DC: Joint Intelligence Military College, 2003): 95-132. Cited hereafter as Moore and Krizan, "Core Competencies."

[213] Kent identified and developed three concepts related to strategic intelligence: Knowledge or what is produced and disseminated; Activity, or how such knowledge is produced and disseminated; and Organization, or how people are grouped to produce and disseminate such knowledge. Kent, *Strategic Intelligence*. Moore and Krizan advocated this approach in their competency work. See Moore and Krizan, "Core Competencies" and David T. Moore, Lisa Krizan, and Elizabeth J. Moore, "Evaluating Intelligence: A Competency-Based Approach," *International Journal of Intelligence and CounterIntelligence*, vol. 18, no. 2 (Summer 2005).

CHAPTER 4
The Shape of Intelligence Sensemaking

Intelligence sensemaking involves a number of overlapping high-level activities. First, intelligence professionals engage in planning or design and then hunt for and gather the materials they require in order to understand issues, answer questions, or explore new ideas. They can be externally motivated by the needs of a customer or they can be self-motivated as a result of an observation, or both. Second, these professionals disaggregate and then reassemble relevant information, trying to determine what it means. At every stage in their work they assess critically their processes and results, seeking to validate both how they are engaged and the outcomes of their engagements. These overlapping activities can be characterized as Planning, Foraging, Marshaling, Understanding, and Communication. They are supported by Questioning and Assessing. Although these elements are discussed separately, it is of course only through their applied interaction that they describe sensemaking.

Planning for Tame and Wicked Intelligence Problems

Making sense of either tame or wicked problems is predicated upon planning. Plans, according to Gary Klein, are "prescriptions or roadmaps for procedures that can be followed to reach some goal, with perhaps some modification based on monitoring outcomes."[214] Creating plans requires "choosing and organizing courses of action on the basis of assumptions about what will happen in the future."[215] Known as *planning*, this process characterizes the "contingencies and interdependencies such as actions that must occur first as a precondition for later actions."[216] When we add the concept that practitioners—through critical thinking—also engage in reflective thinking and learning, both singularly and collaboratively, we may label this process "the art of intelligence design."

---

[214] Gary Klein, "Flexecution as a Paradigm for Replanning, Part 1," *IEEE Intelligent Systems*, vol. 22, no. 5 (September/October 2007), 79. Cited hereafter as Klein, "Flexecution 1."

[215] Klein, "Flexecution 1," 79.

[216] Klein, "Flexecution 1," 79.

With tame problems, where answers and solutions can be anticipated, algorithms can calculate actionable probabilities and repeatedly make sense of the problem.[217] The design of useful algorithms may be complex and they may operate well only in finite and specific environments. On the other hand, how does one plan or design for wicked problems? One answer is to re-imagine the wicked problem as a tame one. However, the repeated occurrence of "unintended consequences" in past scenarios suggests that this is not a good option. Disaggregating what are assumed to be tame problems into their component parts, regardless of the actual problem type, often proves inadequate, as unintended and unforeseen consequences make clear.

Yet planning must occur regardless of problem type. Otherwise, dealing with problems becomes a process of trial and error with no means of agreement on assessing the solutions or even on what are the solutions. Klein considers—along with Rittel and Webber—that planning is an emergent process: Goals are clarified and revised as understanding of the problem grows.[218] He notes,

Goals can be dynamic and can change completely as a function of changing circumstances. Goals can conflict with other goals in ways we can't anticipate or resolve in advance. Goals can carry implications we can't perceive or anticipate until events transpire.[219]

In terms of intelligence creation, this means that larger, strategic goals can—and perhaps must—emerge as sense is made of the problems under scrutiny. Thus, a tasking from an intelligence consumer changes as mindful sense is made of the tasking itself, of the resources that are available for understanding it, and the mix of actors involved.

Klein refers to this reflective problem planning as "flexible execution" or "Flexecution."[220] Within the framework of intelligence sensemaking, it provides a self-reflective process—at the individual and organizational levels—that monitors the goals and whether what is understood or being done is consistent with those goals, modifying those goals as understanding emerges. For example, in examining the situation in Cuba during the summer of 1962, the understanding of the intelligence professionals at the Refugee Processing Center in Miami developed as the summer progressed. A new understanding emerged of what the Soviet forces deployed to the island might be doing.

---

[217] Gary Klein, "Flexecution as a Paradigm for Replanning, Part 2," *IEEE Intelligent Systems*, vol. 22, no. 6 (November/December 2007), 112. Cited hereafter as Klein, "Flexecution 2."
[218] Klein, "Flexecution 1," 81.
[219] Klein, "Flexecution 1," 81.
[220] Klein, "Flexecution 2, 108.

Table 3. Classical Planning and Execution Versus Flexecution		
	Classical Planning/ Execution	Flexecution
End state	Known	Unknown
Preparation	Alternative courses of action Contingencies	Alternative goals and priorities Potential actions
Mode	Increase constraints and control	Fix/flex cycles
Prior plans	Useful	Often obsolete
Commander's intent	Fixed	Continually adjusted Shows goal priorities Shows trade-offs
Strategy	Management by Objectives	Management by Discovery
Accountability	Clear	Unclear

Source: Gary Klein, "Flexecution as a Paradigm for Replanning, Part 2," *IEEE Intelligent Systems*, vol. 22, no. 5 (November/December 2007), 112.

Intelligence planning took place as sense was being made of the situation, (finally) monitoring what they were observing and what it meant. Such deliberations apparently led two analysts to conclude that the indicators for strategic nuclear missiles deployed in Cuba might be valid, resulting in the U-2 overflight of 14 October 1962 and the "discovery" of the missiles.[221] Unfortunately, as has been noted, the intelligence professionals were not sufficiently flexible in their planning to consider that other types of nuclear missiles also might be (and in fact were) deployed on the island.

As summarized in table 3, flexecutive planning seems ideally structured for making sense of wicked problems and taking advantage of the peculiar capability that intelligence has to initiate information actions and then clandestinely determine the adversary's reactions to it. Intelligence professionals are well placed to learn of changes in planning or strategy on the part of adversaries as well as of their own decisionmakers. One can never be certain that all potential planning options are known, but by focusing on "alternative goals and priorities" to which an adversary might migrate, and the potential actions both at the onset of the sensemaking process and as

---

[221] Garthoff, "US Intelligence," 23.

a part of an ongoing, mindful critical reflection, one comes to understand the alternative goals and priorities of one's own actors as well as those of the adversary.[222]

Foraging

Hunting and Gathering

If, as Baumard asserts, "Intelligence, a continuous human activity, gives sense to the stimuli received from the environment [then] these stimuli [must] be passively or actively sought."[223] This requires hunting and gathering. They comprise foraging, which in turn refers to "a wide search over an area in order to obtain something."[224] Whether it describes birds seeking life-sustaining berries, bees scouting for pollen, or people searching for information, foraging describes how animals go about satisfying their needs. For intelligence professionals, foraging describes the means by which the raw materials needed to notice and make sense of phenomena are acquired.

An apt analogy for the foraging activities of intelligence professionals can be drawn from anthropology, where the activities of the hunter-gatherer have been immortalized. In nonagricultural societies people both hunt for specific game and take advantage of what the local environment provides.[225] Similarly, intelligence professionals may seek specific information, often tasking collection systems as part of the search. They also take advantage of existing repositories of information. Neither approach is wholly satisfying nor provides for all of the sensemaker's needs all of the time. However, without the basic act of foraging there can be no sensemaking as there is nothing from which to make sense.

Information foraging is a rich subject about which Peter Pirolli has done extensive work, some of it sponsored by the Intelligence Community's Novel Intelligence from Massive Data research project funded by IARPA's predecessor, ARDA (Advanced Research and Development Activity). Known

---

222 Within the domain of wicked problem consideration there is a means, known as morphological analysis, that may come close to at least identifying all possible planning options. Invented and developed by astrophysicist Fritz Zwicky, the paradigm has been further developed by the Swedish Morphological Society (URL: <www.swemorph.com>, accessed 10 October 2010), and for the U.S. Government by William N. Reynolds, Least Squares Software. Combined with a model of Flexecution, it seems a promising approach for intelligence (re)planning.

223 Baumard, "From Noticing to Making Sense," 31.

224 New Oxford American Dictionary, Apple Computer Edition, 2005, entry under "foraging." Cited hereafter as New Oxford American Dictionary.

225 This remains such a culturally (and perhaps psychologically) essential human activity that the practice remains part of seasonal life in modern post-industrial societies.

as "Information Foraging Theory," Pirolli's work makes use of the optimum foraging strategies of animals as a basis for assessing the human acquisition of information. Underlying information foraging theory is the idea that "humans actively seek, gather, share, and consume information to a degree unapproached by other organisms" and therefore, "when feasible, natural information systems evolve toward stable states that maximize gains of valuable information per unit cost."[226]

Pirolli's research and experimentation devotes special attention to foraging efficiency. Efficiency derives from optimizing the time necessary to achieve a goal, the quality of the achievement, and the satisfaction obtained in doing so.[227] In application to the human information foraging scene, the theory becomes "a rational analysis of the task and information environment that draws on optimal foraging theory from biology and...a production system model of the cognitive structure of [the] task."[228]

Toward a Practice of Intelligence Foraging

Developing an optimal foraging model for information acquisition requires the subject to consider whether to remain at a source that provides a superabundance of information of questionable value, or to seek another, more valuable, source.[229] Pirolli and colleague Stuart Card observe that for human foragers, this involves "a tradeoff among three kinds of processes": exploring, enriching, and exploiting.[230]

These three foraging steps will not seem foreign to traditional intelligence practitioners. "Exploring" is a breadth activity whereby a sensemaker broadly examines a wide variety of information that may or may not be relevant to the issue. The premise is that when one considers a broad variety and volume of data, there is less opportunity to miss "something novel in

---

226 Peter Pirolli and Stuart Card, "Information Foraging," *Psychological Review*, vol. 106, no. 4 (October 1999), 643. Cited hereafter as Pirolli and Card, "Information Foraging."

227 Peter Pirolli, *Information Foraging Theory: Adaptive Interaction with Information* (Oxford, UK: Oxford University Press, 2007), 5. Cited hereafter as Pirolli, *Information Foraging Theory*.

228 Pirolli, *Information Foraging Theory*, 5.

229 In the animal kingdom this quantifiable function is known as the "Conventional Patch Model." It was developed by David W. Stephens and John R. Krebs, *Foraging Theory* (Princeton, NJ: Princeton University Press, 1986); and based on work by Eric L. Charnov, "Optimal Foraging: The Marginal Value Theorem," *Theoretical Population Biology*, 9, no 2 (April 1976): 129-136. Referenced in Pirolli, *Information Foraging Theory*, 7-8.

230 Peter Pirolli and Stuart Card, "The Sensemaking Process and Leverage Points for Analyst Technology as Identified Through Cognitive Task Analysis," paper presented at the 2005 International Conference on Intelligence Analysis, Vienna, Virginia, 2-6 May 2005, URL: <https://analysis.mitre.org/proceedings_agenda.htm#papers>, accessed 11 March 2009. Cited hereafter as Pirolli and Card, "Sensemaking Process."

the data."[231] Speaking in traditional intelligence terms, exploring is like reconnaissance. By contrast, "enriching" is a depth activity. Here the sensemaker identifies areas of interest and focuses attention on those areas. As Pirolli and Card note, this is "a process in which smaller, higher-precision sets of documents are created."[232] Reconnaissance has become more narrowly focused, but highly targeted "surveillance" is not yet in play. Finally, the practitioner "exploits" the results of foraging by thoroughly examining what is found and extracting information as needed. This activity extrapolates from tacit sensemaker behaviors and information-based patterns to create hypotheses about what the information means. At this point, foraging evolves to sensemaking.

The appropriate amount of exploration depends on the context. However, there appears to be a limit after which more information does not increase accuracy although it does increase the sensemaker's overall confidence. One example of this phenomenon was discovered by Paul Slovic in an experiment with odds makers who, when predicting the winners of horse races, were not significantly more accurate if they used 40 variables or only 5 (out of sets of 88 possible variables).[233] What Slovic additionally observed, however, is that the odds makers' confidence did increase directly with the number of variables considered.[234] Slovic's findings about accuracy and confidence are reproduced in Figure 2.

The discussion of how much exploration is needed is germane because in the Pirolli-Card framework the sensemaker may believe she controls the amount of foraging. In a sense this is true. The sensemaker will stop foraging once she believes she has what she needs. But how much information is sufficient? Slovic's results and Heuer's subsequent discussion suggest that practical sufficiency is achieved at lower levels of exploration than expected.[235]

Further, contrary to the beliefs of the sensemaker, it is often information itself that controls the processes. Intelligence professionals are overwhelmed with more and more information that arrives faster and faster and may be valuable for shorter and shorter periods of time. This information flood challenges the sensemaker to efficiently find the information needed in order to make sense of the phenomena or issue under scrutiny in a timely

---

[231] Pirolli and Card, "Sensemaking Process."

[232] Pirolli and Card, "Sensemaking Process."

[233] Paul Slovic, "Behavioral Problems of Adhering to a Decision Policy," Paper presented at the Institute for Quantitative Research in Finance, Napa, CA, May 1973. Cited hereafter as Slovic, "Behavioral Problems."

[234] Slovic, "Behavioral Problems."

[235] See Richards J. Heuer, Jr., *The Psychology of Intelligence Analysis* (Washington, DC: Center for the Study of Intelligence, 1999), 54. Cited hereafter as Heuer, *Psychology.*

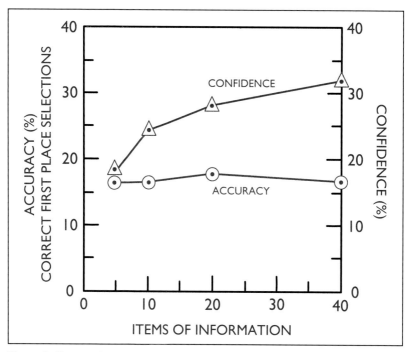

Figure 2. Changes in Predictive Accuracy and Confidence as Available Information Increases. The accuracy of odds-makers' predictions about horse races does not increase as they consider more items of information although their confidence in those predictions does.

Image Source: Richards J. Heuer, Jr., *The Psychology of Intelligence Analysis* (Washington, DC: Center for the Study of Intelligence, 1999), 54. The illustration reproduces one originally presented by Paul Slovic in 1973. See Slovic, "Behavioral Problems."

manner. Does one need to peruse all the information received? Would a different source be more productive by providing more focused information? These are examples of the difficult questions the sensemaker must consider. Compounding her deliberations further is the fact that she must answer this question in foresight; hindsight is too late.

Referring to the animal foraging analogy that underlies information foraging theory, one asks, is the nutrition gained of sufficiently low quality (even though there is lots of it) to justify seeking another source of unknown (but possibly higher) nutritional value? Failure to make the right decision may result in starvation. For the sensemaker this means she may never find what she needs. An implication is that her lack of information may contribute to an error leading to a catastrophic failure. To make such a decision wisely requires something she currently lacks: a practice of foraging. Such a practice puts her, not the information, in control of foraging practice. It provides her with a systematic means of reducing her uncertainty.

Foraging practice begins with an understanding of what it is the sensemaker seeks to know, the foraging resources available, and the urgency of the issue. But how can an intelligence sensemaker know what to look for? To what degree does she need to explore, enrich, or exploit the information? Further, how does she know if she is getting what she needs?

A first step is to think critically about the issue itself and the resources needed. Using a metacognitive, process-focused, critical-thinking model such as that adapted from Richard Paul, Linda Elder, and Gerald Nosich, the practitioner can dissect the issue and her thinking on the issue.[236] She makes assumptions explicit, explores relevant points of view, starts to consider the ramifications of the issue, and she asks important questions about what resources will best inform her about the issue; she considers the context in which she is working, and ponders the alternatives to her reasoning about where and how to forage. This critical thinking defines her foraging activities. She may engage in all three strategies at once or at different times as she engages in the analyses that produce her syntheses and the interpretations necessary to generate knowledge. A well-developed understanding of exploited information may direct her back to do additional exploring or enriching (or both).

In intelligence foraging as it traditionally has been practiced, there is a tendency to linger at a fruitful source rather than to explore elsewhere for the required information. This is a function of the confidence one gains in those systems that previously have supplied information, and the available tools for ferreting out the information. The danger, of course, is that the practitioner may limit the information she acquires and the relevant perspectives it informs. If, for example, the practitioner has a belief that two parties in whom she has an interest communicate via one means and she can acquire technical collection that captures the communications via those means, she may ignore the fact that they also use other methods to communicate. She may subsequently fail to task systems that collect those other communications in the belief that what she is getting suffices. Should the parties suspect that their communications are being targeted, they may engage in deceptive practices over that collected means and use the other non-collected methods for their real exchanges.

---

236 Moore, *Critical Thinking*, especially chapter 1. Also see: Richard W. Paul and Linda Elder, *Critical Thinking: Tools for Taking Charge of Your Professional and Personal Life* (Upper Saddle River, NJ: Prentice Hall, 2002); and Nosich, *Learning to Think Things Through*.

Foraging Challenges

Ongoing research at University College London offers a view of how younger sensemakers likely search for information. The researchers report that people foraging for information spend four to eight minutes viewing each resource.[237] Thus, a great many resources may be consulted but none of them very deeply. Within the context of intelligence, such foraging strategies may facilitate broad searches but leave open the question of whether deeper searches are also accomplished. As Nicholas Carr observes in the July/August 2008 *Atlantic*, Internet search strategies, as epitomized by the operation of search tools such as Google, will put in place a new behavior of foraging and consuming information. Carr notes:

> It is clear that users are not reading online in the traditional sense; indeed there are signs that new forms of "reading" are emerging as users "power browse" horizontally through titles, contents pages and abstracts going for quick wins. It almost seems that they go online to avoid reading in the traditional sense.[238]

From the perspective of actually making sense of issues, these strategies imply that the manner in which issues are understood is also likely to change. Are such new foraging practices suitable to meet the needs of intelligence consumers? Will they be better at satisfying those needs for knowledge than the paradigm previously practiced?

An additional challenge with information foraging is that if the practitioner misses the opportunity to acquire something, it may never again be obtainable. Like the elements of a fleeting interpersonal conversation, the original foraging behavior, if left un-captured, can never be recaptured. Indeed, there may be no indications that such a conversation even occurred.

A further consideration for the practitioner is the "self-marketing" of the information. Vivid stories market themselves much better than do flat ones. Exploited information that supports a favored hypothesis may be preferred over information that does not; an unfortunate reality is that little motivation remains for further exploration. The practitioner is human—she will not likely have a truly agnostic attitude about what she seeks and why.

Compounding this is the idea that sources and means for foraging are self-protective. For example, there is a presumption that sources will continue

---

237 Ian Rowlands and others. "The information Behaviour of the Researcher of the Future (Google Generation)," URL: <http://www.ucl.ac.uk/infostudies/research/ciber/downloads/ggexecutive.pdf>, accessed 27 May 2010.

238 Nicholas Carr, "Is Google Making Us Stupid?" *The Atlantic*, Online Edition (July/August 2008), URL: <http://www.theatlantic.com/magazine/archive/2008/07/is-google-making-us-stupid/6868>, accessed 27 May 2010. Cited hereafter as Carr, "Google."

to communicate via specific means. The methods that capture those communications and the people that support them tend to seek justification. Assets may be kept active after their usefulness expires. The practitioner returns to the same sources over and over because they have been useful in the past and such attention helps keep those sources actively collecting.

Critically assessing what she is doing is one way the sensemaking practitioner may be able to overcome these limiting tendencies. By constantly asking herself how she is thinking about the issue, what she seeks, other perspectives, her assumptions, as well as relevant concepts such as self-deception or adversarial deception, the practitioner may diminish the impact that her preferences play on her foraging decisions.

Another part of this critical assessment is the consideration of the costs of foraging. As Herbert Simon notes,

> What information consumes is rather obvious: it consumes the attention of its recipients. Hence a wealth of information creates a poverty of attention, and a need to allocate that attention efficiently among the overabundance of information sources that might consume it.[239]

Simon in fact anticipates the conclusions of Carr: The superabundance of information available on the Internet (and elsewhere) creates a "poverty of attention" to any one source. Rather, people skim across a great many sources.[240]

Questions the sensemaker must ask include whether or not she can afford to explore an issue's information field further. She needs to consider how certain she is that she has sufficient exploitable information to make sense of the issue. Further, if the issue is inadequately understood, opportunism may affect the foraging as a hunting analogy illustrates. A predator may go out seeking one type of prey and find none of it *but* there may be an abundance of some other kind of game. Within the context of intelligence such opportunism may or may not be appropriate (or even legal) when a technical system or an asset is tasked to provide information for intelligence. Lacking the desired information, a human source might opportunistically substitute what might be perceived as desired or desirable, even if it is not closely related to the issue at hand—or for that matter, even "true." It is used because it satisfices for the immediate term.

---

239 Herbert A. Simon, "Designing Organizations in an Information-Rich World," in Martin Greenberger, ed., *Computers, Communications, and the Public Interest* (Baltimore, MD: Johns Hopkins University Press, 1971), 40-41.

240 Carr, "Google."

Harvesting [241]

A special case of foraging involves "harvested" information. Technical agencies that field systems to gather information also can be characterized by a different model, that of harvesting. The systems employed simply harvest that which lies within their purview, then process it and store it in silos—data repositories from which sensemakers subsequently must forage.[242] Such systems are efficient at creating broad collections; they tend to be inefficient and unreliable when very narrowly focused. Thus, *directed* rather than broad collection of specific phenomena is needed.

Technical collection systems tend to provide—even in the negative—what sensemakers want to find. This can create a potentially dangerous confirmation of an idea that may be invalid. This tendency is exacerbated in the post-Cold War world. Although previously, certain intelligence targets did remain relatively unchanged over considerable periods of time, this is no longer the case.

The harvesting system automatically explores a subset of a larger reality based on a collective agreement among sensemakers that the subset is relevant. The sensemakers subsequently enrich their knowledge and in a best-case situation can exploit the information to provide themselves with the necessary information to make sense of the issue. A challenge is that it is impossible to collect everything. Yet, uncertainty about whether or not one *really* has collected what is necessary remains unresolved. While some technical systems are more usable for "random" exploration than others, most require a preexisting idea about what one seeks. Such preconceptions limit exploration.

Automated retrievals from information repositories typically provide sensemakers with what they believe to be the needed and relevant information—in short, their evidence. The evidence pertaining to specific issues arrives at the sensemaker's desk and reports are issued. At first, the evidence is carefully scrutinized and the system that provides it assessed. As the process repeats, however, as it certainly did in the Cold War era, complacency may set in. Critical assessment of *quality* and *quantity* may cease. And, one day, the evidence that a specific sensemaker requires no longer appears. The reporting on that specific issue withers. If sufficient evidence valuable to *other* sensemakers still is harvested, the value of the continued tasking of the system may

---

241 The author is indebted to Martin Krizan for raising this idea.

242 It should be noted that "reaping" and "threshing" both manipulate the collected information into usable formats. The processed information must be compatible with U.S. law and specific agency procedures for handling such information.

not be questioned. The failure is not noticed and may even be rationalized—perhaps as a failure to forage adequately—*if* it is noticed in the first place. Even when it is noticed, it may be impossible to determine *why* it occurred. Finally, no amount of foraging can discover valuable information if it has not been collected by some system—human or technical—in the first place.

Marshaling

What can be done to revolutionize the way information foraging is accomplished so as to overcome or at least mitigate these problems? Some answers lie in an understanding of marshaling. Part of the sensemaker's practice is to turn foraged and gleaned information into evidence. Doing so requires sifting and other organizing activities to determine which information is relevant to the issue. This is a broad activity, for if the issue has multiple explanations or future possibilities, then evidence will be information relevant to any, many, or even all of those explanations or possible outcomes. In order to make that determination, the sensemaker will need to have identified what those alternatives are and to have collected information (both disconfirming and confirming) about them.[243] This may require foraging from additional resources with all the attendant challenges discussed above. Conceivably this could overwhelm the sensemaker. However, questions such as "if this alternative were true, what would be the evidence for it?" can lead one to identify what she needs to know. Then asking whether or not she sees it, or where she might forage for it (and can do so), starts the process of marshaling. This activity represents a change from traditional practices because the intelligence professional goes beyond what she knows into what she knows she does not know.

Understanding

If we presume that foraging has yielded relevant and valuable information—evidence—on the issue under study, the next step is to determine what it means. This is the heart of sensemaking: evidence is dissected, reassembled

---

[243] Disconfirming alternative explanations (or hypotheses) is one means of avoiding confirming a favored (and possibly wrong) possibility. This disconfirmatory approach is one of the strengths of Richards Heuer and Morgan Jones' independent work with hypothesis testing (Heuer calls it alternative competing hypotheses). How effectively this approach actually works is subject to debate. Another challenge is to define foraging strategies that yield disconfirming (i.e. negative evidence). An absence of evidence is insufficient. See Heuer, *Psychology*, Chapter 8, and Morgan Jones, *The Thinker's Toolkit* (New York, NY: Three Rivers Press, 1995), 178-216. For discussion about whether or not the method effectively disconfirms, see Brant A. Cheikes, Mark J. Brown, Paul E. Lehner, and Leonard Adelman, *Confirmation Bias in Complex Analyses*, Mitre Technical Report, MTR 04B0000017 (Bedford, MA: Mitre, 2004). Cited hereafter as Cheikes and others, *Confirmation Bias*.

with other evidence, and its meanings determined by analyzing, synthesizing, and then interpreting the evidence.

Analyzing

The intelligence issue or question itself requires analytic scrutiny. Different foraging disciplines represent different points of view. In other words, signals intelligence or SIGINT; human sources embodied in HUMINT; images and geospatial data (known as GEOINT), and others tell different stories about a phenomenon. Each story requires dissection. Similarly, within a foraged source, the different perspectives of the sensemaker require dissection. For example, is the sensemaker focusing on individual actors, the actions of a collection of actors, the beliefs that guide the activity, or the processes that determine the actions of the collective?[244]

The disaggregation of each of these perspectives and their associated stories provides a rich brew for sensemaking. For example, in the Cuban Missile Crisis of 1962, such a dissection reveals the intentions and underlying beliefs of the principal actors (Castro, Kennedy, and Khrushchev), as well as the roles and procedures of their larger collectives, in this case the Cuban and Soviet governments (specifically the Soviet Politburo and General Staff) as well as Kennedy's Executive Committee (EXCOM) and indeed the much larger American political collective. However, as rich as these stories become, they remain inadequate in assessing what is likely to happen. Dissecting the evidence is insufficient. Pulling the pieces together becomes the next step.

Synthesizing

Synthesizing is "the combination of ideas to form a theory or system."[245] Even as the intelligence professional analyzes the individual pieces of information, they are synthesized into a mental picture of the larger issue. Pieces of information are implicitly combined even when the sensemaker works within the yield of a particular foraging discipline or within a frame or reference. Such synthesis drives further foraging and analysis.

Synthesis needs to be explicit. In the example developed above, the intelligence professional is required to synthesize the differing trajectories

---

[244] These four groups form the basis of a novel sensemaking approach developed by Monitor 360 for the National Security Agency and incorporated into a course on multi-frame sensemaking. As characterized in the course, the four groups or frames are the "empowered actor," "cooperation and conflict," "beliefs and affiliations," and "roles and procedures." The author is a champion of this approach, and worked with the senior course developer, Bruce Chew, in creating the course.

[245] *New Oxford American Dictionary*, Apple Computer Edition, entry under "synthesis."

of the three principal actors, considering how their beliefs harden or soften their positions and how they are vulnerable to the actions, influences and processes of the groups. Doing so in a systematic fashion leads the intelligence professional to new insights about the situation: what is going on in Cuba and (from the U.S. perspective) what to do about it.

Interpreting

Issues can be dissected and reconstructed in a variety of ways, creating different meanings. Sense must be made of these different meanings. Interpreting, or "the action of explaining the meaning of something," is another component of sensemaking.[246] We may say that whereas analysis and synthesis establish the *what*, interpretation establishes the *so what*.

Depending on the frameworks involved, the interpretations of common information vary widely, as the above referenced experience of London subway bombing victim Rachel North illustrates. An evolving practice within intelligence of establishing competing teams of intelligence professionals to develop different aspects of an issue is an example whereby differing interpretations compete in an effort to establish "ground truth" about issues. A revolutionary approach to sensemaking now being undertaken by analysts from DIA, State, and CIA, is to engage in "adversarial briefing" of principals, where briefers adopt opposing perspectives for a thorough airing of the issue, complete with the participation of the principals themselves.

Communicating

New models of knowledge transfer recognize change in both message and medium. Social networking, peer-reviewed shared multimedia, and interactively blogged communications are examples of these new mediums. The message is short and subject to change by different contributors. Authority is based on consensus. The distinction—if it exists at all—between formal and informal communication is blurred. There are dangers here as authority and truth are no longer necessarily linked. One risk is that the "wisdom of a crowd" can in fact be the "madness of a mob"—a phenomenon occurring in both the public arena and within the IC's blogosphere. In both arenas the loudest voices strive to bludgeon into silence those who would disagree, all the while advancing their egocentric or sociocentric positions. Scientific knowledge and empirical facts matter little in such cases. The danger in both

---

246 *New Oxford American Dictionary*, Apple Computer Edition, entry under "interpretation."

arenas is, of course, that people make poor decisions based on information that later proves to be false.[247]

However, communications within the Intelligence Community remain firmly embedded in traditional formats of printing and briefing. It is true that blogging and informal communications are utilized, but these are used by intelligence practitioners to discuss the issues they make sense of *before* they craft their traditionally formatted assessments, briefings, and reports. The ultimate Intelligence Community report, the President's Daily Brief, remains primarily a hard copy document. While it has been updated incrementally over the years, the presentation remains similar to the summary that was nicknamed "Truman's Newspaper."

As has been noted, newer Intelligence Community directives provide for presenting alternative hypotheses as well as documenting confidence levels in sources and in intelligence assessments. Is this guidance sufficient and valid for sensemaking? In summarizing the "introspective works responding to…intelligence failures," Charles Weiss agrees that intelligence practitioners' failures include a lack of proper attention to hypotheses and data collection efforts that are contrary to what they regard as the most likely interpretation of available information."[248] One danger is that the very judgment about which the sensemaker is least confident might be the one that turns out to be correct. The fallacy of depending on the communication of confidence levels relates to the fact that each assessment or report only fills in some unknown portion of the gaps in the sensemaker's and policymaker's knowledge.

Presenting a novel hypothesis and interpreted argument about which one is uncertain, along with an assessment based on a more likely premise, also carries dangers. A policymaker might disregard the alternate possibility because of the declared lack of confidence (which might stem merely from a lack of evidence) and choose the evidentiarily better-supported hypothesis.[249] Here, a carefully considered, standardized metric of uncertainty could provide one means of assessing and communicating confidence independently

---

[247] The "Swift Boat" controversy of 2004, the matter of autism-caused vaccines, and the uncivil discourse between the major political parties are several non-intelligence examples. These are different from debates, which advance knowledge through open discussion. These are about having one's way. Whoever can shout the longest and loudest wins. See also Peter Miller, *The Smart Swarm: How Understanding Flocks, Schools, and Colonies Can Make Us Better at Communicating, Decision Making, and Getting Things Done* (New York, NY: Penguin Books, 2010), chapter 5. Miller explores the transformation of the crowd to a mob, as well as the self-destructive behavior of the mob, through the example of locusts.

[248] Charles Weiss, "Communicating Uncertainty in Intelligence and Other Professions," *International Journal of Intelligence and CounterIntelligence* vol. 21, no. 1 (Spring 2008), 57-58. Cited hereafter as Weiss, "Communicating Uncertainty."

[249] Weiss, "Communicating Uncertainty," 62.

from the sensemaker. Weiss suggests that either Kent's scale[250] or its more recent instantiation by the Office of the Director of National Intelligence offers an appropriate means by which the uncertainty could be systematically captured.[251] The challenges inherent in such metrics are twofold. First, the evidentiary statistics necessary for their use are "typically unavailable to intelligence analysts—or it they are available, must be based on small samples of past events."[252] Additionally, scoring the conclusions from such small samples across production lines and even from day to day by a single intelligence professional can be observed to be inconsistent. Steve Rieber discusses calibrating sensemakers as a solution.[253] To date no such strategy has been implemented.

Peer-reviewed, discussed and argued findings published in blogs (as a formal means of communicating) would be a novel means of communicating results. If they are readily available to the policy community, they would enhance the capabilities of intelligence practitioners and policymakers alike to collaboratively make sense of the issues that policy faces. Intelligence could truly speak truth to power and policy could speak truth to intelligence. As an alternative communications medium where body language and other non-verbal cues can be read, the adversarial briefing initiative mentioned earlier offers an intermediate solution to the problem of communicating intelligence needs and perspectives.

At the same time, the blogosphere certainly provides an effective locus for discussion of developments surrounding ongoing issues. This is manifest in the IC's implementation of "A-Space" or Analysts' Space, a blogosphere where intelligence professionals can communicate among themselves regarding topical and methodological issues.[254] This new model for managing intelligence practitioners that respects their expertise and capabilities is an integral part of the ongoing revolution in communicating intelligence. A useful future step might be the inclusion of the consumer as part of the discussions; such a step would truly be revolutionary.

---

250 Sherman Kent's estimative probability scale appears in *Sherman Kent and the Board of National Estimates*, Collected Essays edited by Donald P Steury (Washington, DC: Center for the Study of Intelligence, 1994), 137. It is reproduced in modified form in Heuer, *Psychology*, 155.

251 Weiss, "Communicating Uncertainty," 61.

252 Weiss, "Communicating Uncertainty," 60-62.

253 Steven Rieber, "Intelligence Analysis and Judgmental Calibration," *International Journal of Intelligence and CounterIntelligence*, vol. 17, no. 1 (Spring 2004): 97-112.

254 Ben Bain, "A-Space Set to Launch this Month," *Federal Computer Week*, 3 September 2008, URL: <http://fcw.com/articles/2008/09/03/aspace-set-to-launch-this-month.aspx>, accessed 15 June 2010.

In the end, how these actors actually go about understanding the issues they consider is a complex process involving both the foraged and marshaled evidence as well as the beliefs and assumptions of the parties involved. Making sense of this requires both deliberative reasoning and intuitive approaches. Exploring how this works is the subject of the next chapter.

CHAPTER 5
A Practice of Understanding

David T. Moore and Robert R. Hoffman

We begin this chapter by considering intuition, trying carefully to define this nebulous term, and exploring the benefits and hazards of intuitive judgments. In order to accomplish this, an understanding of exactly what is meant by the similarly nebulous term 'judgment' is also needed. We develop our position from the work of Daniel Kahneman and Gary Klein as we explore skill-based and heuristic-based intuitive judgments and how they affect forecasting.[255]

Judgment in intelligence sensemaking, as in a number of other domains, likely improves as one progresses to higher levels of proficiency and expertise. However, prediction is both difficult and inherently unreliable because events of different kinds vary considerably in their inherent predictability. A common view is that prediction has these three features:

- It is the act of forecasting a specific future occurrence,

- It has some associated degree of probability and/or confidence, and

- It is linked logically to one or more specific courses of action that the predictor might take to achieve one or another specific goal.

We believe that another paradigm for shaping how we think about prediction is needed, along with fresh terminology. We present a view centered on the notion of "anticipation of ranges" and the concept of the "course of action envelope" as we begin to envision an alternative model of the phenomenon. We turn next to the implications of intuition and predictability for intelligence sensemaking. We conclude by looking ahead at some preliminary research into how the circumstances for sound reasoning may be improved, and we raise some questions about future directions for the community.

Intuition

It is difficult to wrap appropriate words around the concepts that are at hand, thus care should be taken to make certain distinctions. Sometimes, judgments can be rapid, non-conscious, non-deliberative, and almost seem as if they are immediate perceptions and feelings rather than judgments. These

---

[255] Kahneman and Klein, "Intuitive Expertise."

immediate attitudes can blend seamlessly into more deliberative or conscious reasoning, which in turn can, even if fleetingly, maintain the emotion and immediacy normally associated with intuition. Thereafter, we might say that purely deliberative and non-intuitional thinking might transpire. None of these distinctions is clear-cut in the social science approach to mental phenomena or in the annals of philosophy.

The process of intuitive thinking refers—according to psychologist David Myers—to the ability of individuals to fly "through life mostly on autopilot" even as they function effectively.[256] Neurologist Robert Burton observes this phenomenon in the actions of batters who can hit a ball (or miss it) in advance of their being able to consciously perceive it.[257] Burton also highlights a danger of such non-conscious thinking: People can misinterpret the input, leading us to think twice about relying on intuition for important decisions.[258] After all, as in the batter's case, it seems that intuition-based accuracy is limited to extremely rapid events.

At the same time, in fear and flight behavior, a certain level of inaccuracy can be tolerated. If early hominids misinterpreted, for example, the presence of certain shadows in the tall grass and reacted by running away, and the shadows were in fact benign, then the error had few consequences. Similarly, while one's "gut" may advise one not to get into a particular elevator late at night, people nonetheless tend to board the next one. Is there some stimulus of which one is not consciously aware—perhaps a shadow in the elevator—of someone who might be a mugger or worse? If one listens to one's "gut" and does not get on the elevator car, one will never know. Consequently, one will rationalize, "while I may have been wrong, at least I did not get mugged (or worse)."[259]

Intuitive, or, as it is sometimes called, *automatic* thinking forms the basis for much of our personal sensemaking.[260] It allows us to process

---

256 David G. Myers, *Intuition: its Powers and Perils* (New Haven, CT: Yale University Press, 2002), 16. Cited hereafter as Myers, *Intuition*.

257 Robert A. Burton, *On Being Certain: Believing You Are Right Even When You're Not* (New York, NY: St. Martin's Press, 2008), 75. Cited hereafter as Burton, *On Being Certain*.

258 When and how we intuit accurately (and inaccurately) is one of the themes of this chapter.

259 Only if one gets on the elevator, does not get mugged, travels safely to the desired floor, and then safely exits the elevator does one know that the instinct was incorrect. And even then, one merely has one anecdote's worth of experience contrary to that intuition. That nothing untoward occurred in this *specific instance* is no guarantee of the outcome in any future situations.

260 See Douglas J. Herrmann, and Roger Chaffin, "Memory before Ebbinghaus," in David S. Gorfein and Robert Hoffman, eds., *Memory and Learning: The Ebbinghaus Centennial Conference* (Hillsdale, NJ: Erlbaum, 1982), 35-56 and George A. Miller, "The Magical Number Seven, Plus or Minus Two: Some Limits on Our Capacity for Processing Information," *Psychological Review*, vol. 63 (1956): 81-97. Cited hereafter as Miller, "Limits."

complex inputs that far exceed the "span of immediate apprehension" of approximately seven "chunks" or elements that individuals can consciously process in working (or short-term) memory.[261] Such thinking, for example, explains how people (mostly) drive successfully. Recent research revealing a correlation between cell phone use (especially texting) and accidents leads one to extrapolate that attention-requiring activities such as engaging in professional-level office conversation or dialing and conversing while mobile respectively impede otherwise effective automatic activities such as ten-finger typing or absent-mindedly but safely driving an automobile. In one study sponsored by the British Royal Automobile Club Foundation's Transport Research Laboratory, researchers found that texting while driving contributed to impaired driving more than either moderate alcohol or cannabis consumption.[262] From these and other similar studies it becomes clear that conscious reasoning can distract and thus decrease the effectiveness of "automatic thinking," leading to flawed and sometimes fatal decisions.

With other combinations of activities, people can simultaneously and very successfully perform more than one task. Many people are able, for instance, both to chop vegetables and converse casually. It seems that the mechanics of these two simultaneous activities are more automatic than deliberate. Further, we would expect that driving and conversing emotionally—arguing—would impair one's ability to negotiate traffic, for arguing requires deliberation, whereas the usual affective responses in conversation can be very rapid, or "pre-cognitive." In sum, humans are limited with respect to their mental resources for conscious and non-conscious thinking when engaged in complex tasks.[263] Psychological research has shown that with extensive practice and training one can learn to perform two tasks at once, even apparently incompatible tasks such as taking dictation while reading.[264] On the other hand, in most day-to-day "multitasking," performance generally suffers, even

---

[261] Miller, "Limits."

[262] Nick Reed and R. Robbins, "The Effect of Text Messaging on Driver Behaviour," *Published Project Report PPR 367*, RAC Foundation Transport Research Laboratory, September 2008. URL: <http://www.racfoundation.org/files/textingwhiledrivingreport.pdf>, accessed 9 December 2009. This is but one of a number of similar reports. For a summary of research in this domain see Transportation Research Board of the National Academies, *Selected References on Distracted Driving: 2005-2009*. URL: <http://pubsindex.trb.org/DOCs/Publications%20from%20TRIS%20on%20Distracted%20Driving.pdf>, accessed 9 December 2009.

[263] Given the distracting effect of conscious thought on non-conscious activity in cases such as driving and cell phone use, an interesting experiment would be to examine the distraction posed by non-conscious activities on conscious reasoning.

[264] See Elizabeth S. Spelke, William Hirst, and Ulric Neisser, "Skills of Divided Attention," *Cognition*, vol. 4 (1976), 215-230; and Christopher A. Monk, J. Gregory Trafton, J. G., and Deborah A. Boehm-Davis, "The Effect of Interruption Duration and Demand on Resuming Suspended Goals," *Journal of Experimental Psychology: Applied*, vol. 14 (December 2008): 299-313.

for individuals of the Web generation who are widely believed to be skilled at multitasking.[265]

There is another side of intuitive reasoning that sometimes works in opposition to one's survival. Intuitively reasoned responses to stress are often highly focused and narrow. Laurence Gonzales notes that in such reactions "the amygdala…in concert with numerous other structures in the brain and body, help to trigger a staggeringly complex sequence of events, all aimed at producing a behavior to promote survival."[266] But, in many cases, behavior is also locked down. Citing the cases of Navy and Marine Corps pilots who fly their aircraft into the "round down" or stern of aircraft carriers as well as his own experience of nearly crashing while landing a private plane, Gonzales observed that other vital input becomes "irrelevant noise, efficiently screened out by the brain."[267] However, unless such input is accounted for, accidents happen and people die. Non-intuitive input, then, needs to be considered; however, it may take too long to accomplish.

Intuitive or automatic thinking is a survival mechanism.[268] Gonzales notes that such mechanisms "work across a large number of trials to keep the species alive. The individual may live or die."[269] But over time—and in reference to humans—generally more live than die, leading to evolution and the genetic transmission of the "reflex." If a particular set of behaviors confers a survival value, that set can become more widespread in the population. Seen in this light, unease at entering an elevator at night could be a modern instance of sensing shadows in the tall grass.

On a shorter time horizon, people use experience-based intuitive patterns or mental models. These patterns or models direct how situations are perceived and how they are responded to. Mental models provide

---

265 For recent work in this area see Paul Raeburn, "Multitasking May Not Mean Higher Productivity," NPR News: *Science Friday*, 28 August 2009, URL: <http://www.npr.org/templates/story/story.php/storyId=112334449&ft=1&f=1007>, accessed 21 February 2010. Raeburn interviews sociologist Clifford Nass about his work on multitasking. For more on Nass' work see Eyal Ophir, Clifford Nass, and Anthony D. Wagner, "Cognitive Control in Media Multi-taskers," *Proceedings of the National Academies of Science, Early Edition*, vol. 106, no. 37, 15583-15587, URL: <http://www.pnas.org/content/106/37/15583.full>, accessed 18 March 2010 ; and Lin Lin, "Breadth-Biased Versus Focused Cognitive Control in Media Multitasking Behaviors," *Proceedings of the National Academies of Science*, vol. 106, no. 37 (15 September 2009): 15521-15522, URL: <www.pnas.org!cgi!doi!10.1073!pnas.0908642106 PNAS>, accessed 18 March 2010.

266 Laurence Gonzales, *Deep Survival: Who Lives, Who Dies, and Why* (New York, NY: W.W. Norton, 2003), 35-36. Cited hereafter as Gonzales, *Deep Survival*.

267 Gonzales, *Deep Survival*, 39.

268 See William James, *Principles of Psychology*, two vols. (New York, NY: Henry Holt and Company, 1890).

269 Gonzales, *Deep Survival*, 39.

rapid shortcuts to determining the nature of situations and the appropriate responses.[270]

The challenge is that some intuitive mental models can ill-serve the individual. They can be incomplete or can contain incorrect concepts. As Gonzales notes, they can impede our "ability to use working memory properly, to process new information from the world, and to integrate it with long-term memory. And there's plenty of evidence that while they're not always lethal, such lapses are perfectly normal and happen all the time."[271]

In a recent article, Daniel Kahneman and Gary Klein engage in a surprising collaboration to clarify the roles of intuition in human decisionmaking, particularly in the realms of expert judgment.[272] Their collaboration is surprising because Kahneman and Klein come from different communities of practice. That Kahneman, as a recognized leader in the "judgment under uncertainty" decisionmaking community, and Klein, as a recognized leader in the naturalistic decisionmaking movement, find they agree on intuitive expertise opens new partnering opportunities for adherents of both paradigms to seek other common ground, enhancing our understanding of decisionmaking and judgment.

Their shared concern is with how "skilled intuitive judgment develops with experience" and the nature of "the activities in which experience is more likely to produce overconfidence than genuine skill."[273] The authors note that the "judgments and decisions we are most likely to call intuitive come to mind on their own without evoking cues and of course without an explicit evaluation of the validity of those cues."[274] Intuition and intelligence sensemaking are linked by the very concept of judgment as developed by Kahneman and Klein.

We turn next to an examination of types of judgment, distinguishing between those that are skill-based and those that rely on "heuristics," or learning through personal discovery whether rules of thumb are valid shortcuts to understanding an issue. We then link this dissection of judgment to the work of intelligence professionals.

---

270 Gary Klein and Robert R. Hoffman, "Macrocognition, Mental Models, and Cognitive Task Analysis Methodology," in Jan Maarten Schraagen, Laura Grace Militello, Tom Ormerod and Raanan Lipshitz, eds., *Naturalistic Decision Making and Macrocognition* (Aldershot, UK: Ashgate Publishing Limited, 2008), 57-80.

271 Gonzales, *Deep Survival*, 79.

272 Kahneman and Klein, "Intuitive Expertise."

273 Kahneman and Klein, "Intuitive Expertise," 515.

274 Kahneman and Klein, "Intuitive Expertise," 519.

Types of Judgment

First we must clarify the meaning of "judgment." A judgment can be an observer's belief, evaluation or conclusion *about* anything—one can form a judgment about anything of interest, including one's own reasoning.[275] Judgment also describes a *process*, surely more than one kind of mental process, by which one reaches a decision. Judgment can be expressed as affective evaluation (example: That is a good thing), objective evaluation (It looks like a cat, but it is just a stuffed cat), or categorical assignment (My judgment is that this is a case of highway robbery). Judgment as process can also be described as apodictic, modal, or oral, among others.[276]

Kahneman and Klein relate judgment to intuition with these diverse examples:

> The firefighter feels that the house is very dangerous, the nurse feels that an infant is ill, and the chess master immediately sees a promising move. Intuitive skills are not restricted to professionals: Anyone can recognize tension or fatigue in a familiar voice on the phone.[277]

A standard approach to understanding a phenomenon invokes categorization. For example, judgment can be seen as taking one of two forms: skill-based intuitive judgment, and heuristic-based intuitive judgment.

Skill-Based Intuitive Judgments

A simple example of skill-based intuitive judgment is the act of parallel parking by an experienced driver. The experienced driver has repeatedly parked in a similar manner over years of operating a vehicle and has become familiar with its dimensions; also, parking spaces are fairly uniform in size. These factors lead to a likely successful conclusion—dent-free parking.

Examples of skill-based intuitive judgment also seem to occur routinely in sports where repetition facilitates learning. A particular move becomes a habit, a reflex, or as we would say, automatic. The author (Moore) has repeatedly and deliberately capsized and righted his kayak in a similar manner over years of paddling on the Chesapeake Bay, nearby rivers, and elsewhere. He is intuitively familiar with his boat, how it behaves under various circumstances, and what it can be made to do (its "affordances"). He is also familiar with the supporting capacity of various paddles, and the motions

---

[275] See James McCosh, LL.D., *Intuitions of the Mind: Inductively Investigated* (London: UK: Macmillan and Company, 1882).

[276] See Franz Brentano, *Psychology from an Empirical Standpoint*, Antos C. Rancurello, trans. (New York, NY: Humanities Press, 1973). Originally published as *Psychologie vom empirischen Standpunkte*, by Dunker and Humblot, Leipzig, Germany, 1874.

[277] Kahneman and Klein, "Intuitive Expertise," 519.

of his body and paddle. The upshot of such skill-based practice is that in an emergency, the author does not have to deliberate about what to do. Instead, he *knows* how to position himself and "roll up," righting the kayak. In actuality, intuition allows him to paddle in such a manner that post-capsize action is typically unnecessary. He knows the things he can do with his own body, his paddle, and the kayak to prevent capsize—his own "effectivities."[278]

Successful skill-based intuitive judgment presumes "high-validity environments and an opportunity to learn them," such as those found in parallel parking and kayak rolling.[279] This process only works when the environment surrounding the issue "provides adequately valid cues," and the sensemaker "has time to learn those cues."[280] Kahneman and Klein note that in intuition of this sort, "[no] magic is involved. A crucial conclusion emerges: Skilled intuitions will only develop in an environment of sufficient regularity, which provides valid cues to the situation."[281]

In terms of intelligence sensemaking, successful intuitive judgment arises from the tacit knowledge of experts who assess "normal" (in Kuhnian terms) situations, or as has been discussed above, the tame, familiar or regularly occurring kinds of problems (although they may be quite complex). Situations involving state-based actors or others for whom events tend to follow from earlier, observable indicators are an example of environments suitable for the operation of skill-based, expert intuitive judgment.

Heuristic-Based Intuitive Judgments

Heuristic-based intuition relies on "rules of thumb" in order to make sense of situations.[282] Rather than making judgments based on deliberation, the experienced sensemaker can at times recognize when a case fits a nominal type; that is, it is similar in many respects to a well-known instance of the situation. Based on recognition that a given case fits a type, the practitioner knows immediately what principles to apply or actions to take. Heuristic-based intuition, as Klein and Kahneman define it, is rapid and automatic,

---

278 Michael Young, Yi Guan, John Toman, Andy DePalma, and Elena Znamenskaia, "Agent as Detector: An Ecological Psychology Perspective on Learning by Perceiving-Acting Systems," in Barry J. Fishman & Samuel F. O'Connor-Divelbiss, eds., *Fourth International Conference of the Learning Sciences* (Mahwah, NJ: Erlbaum, 2000), 299.

279 Kahneman and Klein, "Intuitive Expertise," 519.

280 Kahneman and Klein, "Intuitive Expertise," 520.

281 Kahneman and Klein, "Intuitive Expertise," 520.

282 The term "rule of thumb" has unclear origins in antiquity, likely related to the idea of using the thumb as a measurement device. In computer science and cognitive science, heuristic rules are distinguished from algorithms. The latter are known to give precise answers in finite time. A heuristic is any short cut that is not guaranteed to give precise answers, but takes less time than is required to use an algorithm.

almost like skill-based intuition. There is no pause to reason through a decision or judgment, but there is an act of recognition. The sensemaker recognizes that actions learned or developed in one context—rules of thumb—are appropriate in another situation. Where the environment is sufficiently unstable so as to preclude recognition and learning of appropriate cues (or the time to do so is too short), heuristics operate similarly to make sense of the situation and offer solutions. Often accompanying such judgment is an "unjustified sense of confidence," a concept Kahneman terms the "illusion of validity."[283] Under such circumstances, the success of intuitive judgment in such conditions may be limited.

However, it should be noted that the combination of skill-based and heuristic-based intuition confers a benefit to mindful experts: a sense of when a case seems typical at first glance, yet there is something not quite right about it. While the less experienced person may be lulled into believing the case fits a certain type, expert decision makers react differently. They note some worrisome clue that raises questions. "Maybe this is not a typical case," they venture. Eventually they may come to see that the case is in fact atypical. So informed, they make a different judgment, sometimes in disagreement with other experts. As we will now see, intuition certainly becomes a part of the intelligence process when practitioners make, or fail to make, useful and accurate predictions.

The Question of Predictability

Some in the IC argue that the pertinent aspects of all intelligence problems can be adduced, "if we only knew more" or "had the right algorithm or method." However, we mislead ourselves if we believe that any messy problem can be resolved with a probability-juggling program. The authors are reminded of the observation, "There are the hard sciences and then there are the difficult sciences."[284] It is both impossible and inappropriate to attempt to remake social and cognitive sciences into emulations of calculational physical sciences. If the reduction were possible, someone would have achieved it, or would have at least made demonstrable progress toward its realization, in the 200-plus years during which psychology and the other "social sciences" have called themselves "sciences." If reduction was appropriate, and we could get "the right information to the right person at the right time," we would not need that right person—"truth" would be self-evident.

By contrast, sensemaking is all about creating and recognizing proximate knowledge and meaning through human cognition. A useful question is

---

283 Kahneman and Klein, "Intuitive Expertise," 517.
284 Attributed to anthropologist Gregory Bateson.

not "How can cognitive work be automated?" but "In what ways and to what extent might technology and software amplify and extend the human ability to engage in cognitive work?" In other words, the unexpected happens, even on a large scale, as is illustrated by the inaccurate predictions offered by a number of books published in the 1995-2001 period that addressed the issues of the looming 21st Century and missed climate change—and more importantly from the U.S. perspective—the likelihood of attacks such as those by Al Qaeda on the World Trade Center and the Pentagon on 11 September 2001.[285]

The IC's flagship, publicly available forecasting reports, the National Intelligence Council's (NIC) quinquennial "global trends" reports, also missed the opportunity to bring attention to a "global war on terror," although effects of climate change did receive consideration in the report published in 2000.[286] Inclusion of this latter driver is not surprising considering that the NIC's brief examination of climate change occurred at least a decade into the discussion and debate about "global warming," suggesting that the phenomenon was known even if the impact was not clearly predictable. One could argue that the likelihood of a war on terror was also knowable. At least one intelligence adviser, Richard Clarke, warned of what would come to be known as the 9/11 attacks during the spring and summer of 2001; that is, prior to the events of that tragic day.[287] The fact that some people did anticipate these trends and events relates to a conclusion of Kahneman and Klein, namely, that "widely shared patterns of association exist, which everyone can recognize although few can find them without prompting."[288]

Interestingly, *The Atlantic Monthly* did warn in a speculative piece published during the summer of 2005 of a looming economic crisis.[289] It blamed the crisis in part on people using home equity to buy stocks—at least getting "right" the fact that the then-future 2009-2010 crisis was (at least in part) about real estate.[290] The real culprit—as has become clear—was the

---

[285] Peter Schwartz's *The Art of the Long View* (New York, NY: Doubleday, 1996) is an example. An irony here is that previously Schwartz headed the forecasting (scenario planning) unit at Royal Dutch Shell, which was (at least according to Schwartz) often correct in their predictions. Schwartz, conversation with the author, September 2006.

[286] National Intelligence Council, *Global Trends 2015: A Dialogue About the Future with Nongovernmental Experts* (Washington, DC: Government Printing Office, 2000), URL: <http://www.dni.gov/nic/NIC_globaltrend2015.html>, accessed 15 February 2010. It should be noted that climate change was already a discussed topic at the time.

[287] As also did others according to Berkowitz (referenced above). While a failure to alert and warn was part of the tragedy, so was a failure to *believe and act*.

[288] Kahneman and Klein, "Intuitive Expertise," 520.

[289] See James Fallows, "Countdown to a Meltdown," *The Atlantic Monthly*, July/August 2005, URL: <http://www.theatlantic.com/doc/200507/fallows>, accessed 15 February 2010.

[290] As seen from hindsight. Mr. Fallows had no way of knowing in foresight which parts of his speculative piece would come "true."

behavior of banking institutions, not individual investors. In both cases greed was a motivating factor.[291] However, the same *Atlantic* piece also speculated on the impending death of Fidel Castro, expected in 2008.[292]

Some examples of mindful, heuristic-based decision making, especially pertinent because they involve the thinking habits of senior U.S. civilian and military officials as well as of their strategic advisors, are discussed in Neustadt and May's *Thinking in Time.*[293] The authors point out that a sensemaker's awareness of historical decision making in even loosely analogous situations helps to keep at bay the further unsettling idea that the present circumstances constitute a "crisis." Even in the absence of politically or militarily identical precedents as a guide, they note that George Marshall, Chief of Staff of the U.S. Army, read history and possessed "the kind of mental quality that readily connects discrete phenomena over time and repeatedly checks connections."[294] Decisions, as informed at times by the IC, might benefit from Neustadt and May's recommendation that heuristic-based decisions can grow from "imagining the future as it may be when it becomes the past."[295] This advice would seem to be useful in making decisions on a strategic basis rather than a more tactical, reactionary plane. From a national intelligence perspective, where proactive decisionmaking in the face of impending change in the national security environment ought to reign supreme, Neustadt and May's heuristic advice about "thinking in time" becomes apt: "You may even [achieve] headway toward identifying changes that might be *made* to take place."[296]

However, like all strategies for framing and re-framing, caution is needed, perhaps especially in considering Neustadt and May's specific method of heuristic reasoning. An awareness that some current occasion or event repeats one that occurred in the past, or that the patterns bear some similarities or symmetries, can serve as either a flashlight or a blindfold. The pattern suggests where to look in finding evidence, but it might also prevent one from looking for other possibilities and particularly for the unique elements of the situation at hand. The characterization by some of the 1991 Gulf War as "another Vietnam" cognitively blinded them to the specifics of the

---

291 But greed, although a constant or enabling condition, is not a good explanation of cause.

292 As of March 2012, Castro still lives.

293 Richard E. Neustadt and Ernest R. May, *Thinking in Time: The Uses of History for Decision Makers* (New York: The Free Press, 1986). See especially chapters 13 and 14. Cited hereafter as Neustadt and May, *Thinking.*

294 Neustadt and May, *Thinking,* 252.

295 Neustadt and May, *Thinking,* 253-254.

296 Neustadt and May, *Thinking,* 259.

case at hand. Similarly, those who considered the second Gulf War as one that completed the first apparently missed key unique elements of that situation. Even so, there are instances when such reasoning is possible and valuable, particularly when expertise can be acquired and is useful.[297]

The two types of intuition suggested here—skill-based and heuristic-based—arise from a number of cognitive processes, including "the operations of memory."[298] While Kahneman and Klein disagree on the frequency and utility of non-skill-based intuitive judgment, they do agree that intuitive judgment may be limited in situations when the decisionmaker is either unskilled or the environment fluctuates irregularly and therefore cannot be learned. "Anchoring," which is the biasing of a judgment because of the framing of the initial question, and "attribute substitution," arising from replacing a difficult question with an easier one, are two contributors to such flawed intuitive judgments.[299]

If venturing predictions is too uncertain for intelligence professionals, what can be done? We address this in our next section as we consider in more detail both why intuitive predictions fail, and when they may succeed. We also present an alternative paradigm for assessing the future.

Thinking About Anticipating

Jurisprudence, clinical psychology, and economic forecasting are all examples of domains where accurate prediction is difficult or impossible, and it is not terribly clear what it means for a person to be an expert in any of those fields. In the realm of jurisprudence, studies of the low rate of successful intuitive predictions about future recidivism among paroled offenders serves as one of many pieces of evidence showing that even highly experienced professionals in certain domains may be no better than laypersons at making intuitive judgments. As recounted by Myers, a 1998 Canadian Solicitor General research team found that the experts—in this case the clinicians—were one of the "*least* accurate predictors of future criminality."[300] Researchers at the University of Minnesota found such "expert" predictions were on average 10 percent less accurate than ones made with "mechanical," i.e. statistical, actuarial, or algorithmic techniques. However, they did

---

297 We discuss this below in more detail along with a means of making use of the past to better understand the present and explore the future.

298 Kahneman and Klein, "Intuitive Expertise," 521.

299 Kahneman and Klein, "Intuitive Expertise," 520-21.

300 Myers, *Intuition*, 173; emphasis in original.

observe that in 63 of the 134 cases they studied, clinical predictions fared as well as mechanical ones.[301]

Admittedly one explanation (at least in the latter case) is that lack of timely feedback inhibits the acquisition of the skills needed for successful prediction. Psychologist James Shanteau observed that domains where the practitioner's task is to predict human activity have few "experts."[302] Indeed, the expertise may lie in tasks other than those that are assumed to be the principal task goals. Thus, for instance, we might think that to be an expert clinical psychologist, one would have to be able to correctly diagnose and then correctly predict the likelihood of recovery, and then correctly treat the client, but the true skill might lie in being a very good listener. Human activity often fails to provide the needed cues for timely feedback, and at a collective level is subject to too many unpredictable events and decisions at the same time as it is subject to known trends and forces. The viral spread of unsubstantiated ideas on the World Wide Web through such mechanisms as YouTube offers an example of how weak feedback at the collective level breeds low intrinsic predictability.

These domains contrast with those where it is relatively easy to identify experts on the basis of operational definitions that specify what it means for a person to be an expert, in terms of performance at principal task goals. Examples would be medicine, weather forecasting, piloting, musical performance, and industrial process control. These latter domains are known by convenience as "Type 1." They contrast with the formerly described domains, known as "Type 2."

How professionals fare in estimative judgments in these domains has been studied. The results of one study of members of the intelligence unit of the Canadian Privy Council Office, carried out by Defence Research and Development Canada, concluded that with ample time and resources the "quality of analytic judgments was 'good to excellent' across [a series of] various indices."[303] The researchers noted, "[experts] that perform well tend to work in areas that provide timely, unambiguous feedback on their performance"—in other words, in stable, Type 1 domains.

---

301 William M. Grove, David H. Zald, Boyd S. Lebow, Beth E. Snitz, and Chad Nelson, "Clinical Versus Mechanical Prediction: A Meta-Analysis," *Psychological Assessment*, vol. 12, no. 1 (2000), 25. Cited hereafter as Grove and others, "Meta-Analysis." As recounted by Myers (p. 173) 63 cases were also a draw; in only 8 cases did clinical methods fare better.

302 James Shanteau, "Competence in Experts: The Role of Task Characteristics," *Organizational Behavior and Human Decision Processes*, vol. 53 (1992): 252-266.

303 David R. Mandel, Alan Barnes, and John Hannigan, "A Calibration Study of an Intelligence Assessment Division," PowerPoint Presentation, Defense Research and Development Canada, Toronto, CA, n.d. Cited hereafter as Mandel, Barnes, and Hannigan, "Calibration Study."

For Type 2 domains, however, the evidence points the other way: we can expect intelligence professionals to be poorer at point prediction unless the predictions are very near term. This tendency is confirmed by the NIC long-range predictions contained in the *Global Trends* papers. In a related example, *The Economist* found greater accuracy in their control group than among the experts in a set of economic point predictions spanning 1984 to 1994.[304] About this experiment, the members of the U.S. Commission on National Security observed that of four groups of people—"finance ministers, chairmen of multinational corporations, Oxford University economics students, and, as a control group, four London garbage collectors"—"[every] group did poorly; the garbage collectors, as it happened, turned out to be the most accurate."[305] In commenting about its own study, *The Economist* concluded, "The contents of dustbins could well be a useful leading economic indicator."[306]

A Space-Time Envelope of Anticipation

What actually occurs when we think about an event or trend lying in the future? Hoffman suggests that causality arises from what he refers to as long-term forces and abstractions, rather than near-term events and decisions. We project past events, decisions, and perceived forces and abstractions forward in space and time toward the present effect or event as we begin trying to make sense of the future. Then we envision the reach and consequence of that effect or event further, into the future; we extrapolate future forces and abstractions from ongoing events and decisions. We therefore convert events of the present moment into a "specious present."

This all occurs in foresight. But we subsequently look at events in backward fashion. Someone got onto an airliner with a bomb in their underwear. What explains this? We can reason backward across actual time from effects and aftereffects (the attempt by the bomber to detonate his explosives and the successful preventative intervention) to causal explanations. And we

---

304 The United States Commission on National Security, *New World Coming: American Security in the 21st Century*, Study Addendum, Phase 1 (July 1998-August 1999), 1. Cited hereafter as Commission on National Security. Since the National Intelligence Council has an "Economics" portfolio, one safely can surmise that economics is an intelligence issue. Therefore the results from the experiment by *The Economist* study has relevance to the practice of intelligence.

305 Commission on National Security, 1. One might well wonder how a group of four intelligence professionals might have fared. Hoffman suspects that the researchers probably only looked at prediction hit-rate and did not drill down into the reasoning and knowledge content relied upon. It is there, he suspects, that the expertise of the "experts" would have emerged. Still, because economics is a Type 2 domain, the results are not implausible.

306 *The Economist*, "Garbage In, Garbage Out," (U.S. Edition), 3 June 1995, 70.

always find those causal explanations; we're certain they're the right ones. For Type 1 domains we may be good at this. The very nature of Type 2 domains precludes any reasonable expectation of accuracy on our part about causal explanations. But, as Kahneman and others point out, people are nonetheless certain they are right: the "illusion of validity" prevails. Such cogitation in hindsight is a quite rational kind of sensemaking, as it is the reconsideration of a case after one has obtained more information about it. It works because we only consider what we knew afterward and what we could or should have known afterward. We do not consider the event from the perspective of what we knew *before* the event occurred, and this feeds into the blame game.

Such human thinking may be illustrated as a series of interlocking triangles set across a "space-time envelope of indeterminate causality" as shown in figure 3. "Effects" or events are predicated upon other events and decisions that are themselves influenced by forces and abstractions. They in turn give rise to new (or continuing) events and decisions that are in turn embodied in

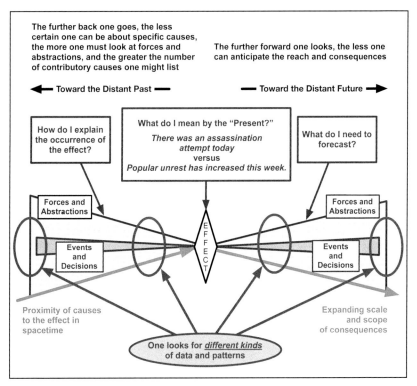

Figure 3. Space-Time Envelope of Indeterminate Causality.
Source: Robert Hoffman.

new (or continuing) forces and abstractions. These predicate and subsequent factors can be viewed from a variety of scales. Using a model of weather forecasting (from which this "envelope" was developed), we observe that at each scale, there are particular dynamics having particular space-time parameters. As one progresses from higher scales to lower, each scale provides the "boundary conditions" or forcing events for the scale below it. Each scale requires the examination of different kinds of data and emergent patterns as well as differing meanings. For example, at a microscale in weather forecasting, one might simply look up and observe what is going on locally over a period of a few minutes; at a continental or "synoptic" scale one makes sense of radar or satellite images showing events that transpire over a period of hours; at global scale observations consider the actions of the jet stream and the worldwide effects of phenomena such as *El Niño* over time spans of months. Dynamics are coupled to space-time, and because time is a factor, there is an explicit expression of increasing uncertainty the farther one is from the effect or event—either in the past or in the future. Placing an episode of intelligence sensemaking into this framework might help keep the thinking honest.

Let us consider that the point of view of the observer is one of an "unspecific present." This is not a single point in time but a range of time, a little in the future and a little in the past, depending on the context or events under analysis. Hoffman's model allows for the factors—at differing scales—to be mapped out conceptually as is shown in figure 4.

At the conclusion of making such a diagram, one has mapped out a path of reasoning—forward in foresight, backward in hindsight—identified the drivers, and placed them in one of three space-time envelopes: past, present, or future. The creation of diagrams of this sort has been useful in charting analyst reasoning and knowledge, making it explicit for discussion and also contributing to knowledge capture and preservation.[307] The diagram facilitates sensemaking by making explicit what one knows and when one knows it as well as how the various trifles of information are related and contribute to one another.

---

[307] Robert R. Hoffman, "Use of Concept Mapping and the Critical Decision Method to Support Human-Centered Computing for the Intelligence Community." Report to the Palo Alto Research Center (PARC) on the Project, "Theory and Design of Adaptable Human Information Interaction Systems for Intelligence Work," Novel Intelligence From Massive Data (NIMD) R&D Program, Advanced Research and Development Activity (ARDA), Department of Defense, Washington, DC, November 2003; and Robert R. Hoffman, "Making Good Decisions about Decision-aiding?" panel on "Beyond Requirements: Decision Making Developing Software Tools for Intelligence Analysts" 49th Annual Meeting of the Human Factors and Ergonomics Society, Orlando, Florida, 2005.

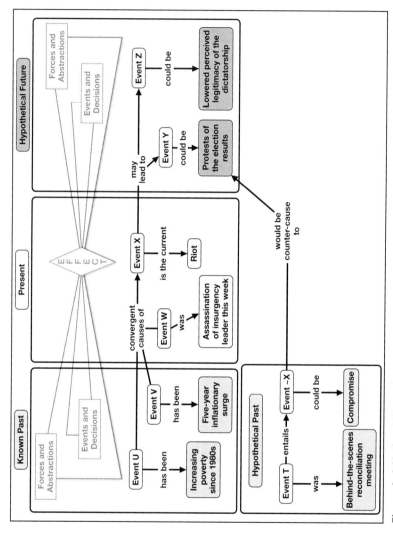

Figure 4. A Notional Example of Causality Across the Space-Time Envelope.
Source: Robert Hoffman.

84

Anticipating Deception: Applying the Space-Time Envelope

A recurring worry within alert intelligence services is whether they are being deceived by their adversaries. From Troy, in the second millennium BCE, through to mis-direction schemes in World War II, and on to the lead-up to the Iraq invasion of 2003, adversarial deception has played a strong or decisive role in final outcomes.[308]

World War II's Operation Mincemeat is of particular interest for the opportunity it affords for counterfactual examination, and thus for a sensemaker's anticipation of future strategic deception.[309] The premise of the Mincemeat deception was to employ a stratagem known as the "lost dispatch case," a sort of portable Trojan Horse, to mislead the Germans about where the Allies would invade as they moved from North Africa onto the European continent. To both sides, Sicily was the logical candidate for the initial landings. Thus, the English proposed floating a corpse ashore at Huelva, Spain, with a dispatch case containing letters and other references to invasion points through the Peloponnese, and Sardinia and Corfu; Sicily was to be a diversion. As both Ewen Montagu and Ben Macintyre make clear, the Germans decided the recovered documents were authentic and redeployed troops to strengthen their positions in the incorrect locations. Subsequently the Allies faced considerably less resistance when they landed on Sicily.[310]

However, could the Germans have come to a different conclusion? Such counterfactual questions normally leave us with a less than satisfying answer. But in this case, in light of first-hand and well-founded knowledge of decisionmaking on both sides, we can marshal the authentic beliefs, events, and drivers that led to a decision to make a well-informed estimate of what could have evoked an opposing decision. As shown in figure 5, key evidence and beliefs, when laid out in the space-time envelope of anticipation, would facilitate the counterfactual process by indicating what underlies the German belief that the documents were authentic. Behind each "event" are in fact chains of evidence that could be explicitly included in a more detailed

---

308 For a very interesting take on the many layers of Iraqi deception see Kevin Woods, James Lacey and Williamson Murray, "Saddam's Delusions: The View From the Inside," *Foreign Affairs* (May/June 2006). It seems that Hussein was himself deceived even as he strove to convince his neighbors that he had weapons of mass destruction (WMD) and strove to convince the United States that he had gotten rid of his WMD—which apparently was the ground truth.

309 See Ewen Montagu, *The Man Who Never Was: World War II's Boldest Counterintelligence Operation* (Annapolis, MD: Naval Institute Press, 2001, reprinted from 1953 original). Cited hereafter as Montagu, *The Man Who Never Was*. The story, a first-hand account by Montagu as one of its planners, is also told in the 1956 movie of the same name. Also see Ben Macintyre, *Operation Mincemeat* (New York, NY: Harmony Books, 2010). Cited hereafter as Macintyre, *Operation Mincemeat*.

310 Macintyre, *Operation Mincemeat*, 284-285; Montagu, *The Man Who Never Was*, 139-146.

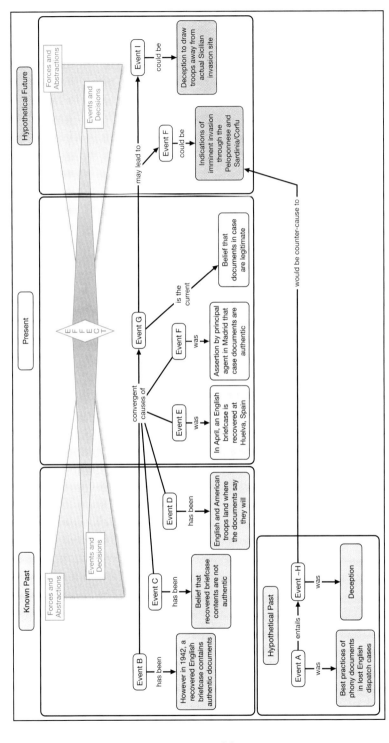

Figure 5. A Space-Time Envelope of Anticipation Regarding the Authenticity of the Documents Recovered at Huelva, Spain, 1943.
Source: Authors.

version of this sensemaking approach.[311] Had the Germans employed this cognitive tool, a different decision regarding the authenticity of the documents and therefore the location of the invasion *might* have been reached. The Allies might have been repulsed and the events of the remainder of the War would have been different.[312]

Implications of Visualizing Anticipation

Diagramming using Concept Maps (and related kinds of diagrams called causal maps and cognitive maps) has been used as a de-biasing technique for analysis under uncertainty. This use is well known in the field of business and strategic management:

> Causal maps allow the map maker to focus on action—for example, how the respondent explains the current situation in terms of previous events, and what changes he or she expects in the future. This kind of cognitive map is currently the most popular mapping method in organization theory and strategic management.[313]

And in the pages of *The Journal of Strategic Management*, Gerard Hodgkinson and his colleagues added:

> In addition to providing a useful means for gaining insights into the nature and significance of cognitive processes underpinning strategic decision making, this dynamic emphasis on antecedents, behaviors and consequences, renders causal cognitive mapping techniques particularly attractive as a potential means for overcoming the effects of framing (and possibly other cognitive biases) in situations involving relatively complex decision scenarios.[314]

Hodgkinson *et alia* investigated "the extent to which judgmental biases arising from the framing of risky decision problems [could] indeed be eliminated

---

[311] This is known as "inference diagramming" and was developed for jurisprudence by jurist John Henry Wigmore (1863-1943). David Schum developed the paradigm for intelligence work, and Moore explored applications of it for intelligence work. See John Henry Wigmore, *The Science of Proof: As Given by Logic, Psychology and General Experience and Illustrated in Judicial Trials*, 3rd edition (Boston, MA: Little, Brown,1937); David A. Schum, *Evidence and Inference for the Intelligence Analyst*, Two Volumes, (Lanham, MD: University Press of America, 1987); and David T. Moore, *Creating Intelligence: Evidence and Inference in the Analysis Process*, MSSI Thesis chaired by Francis J. Hughes (Washington, DC: Joint Military Intelligence College, July 2002).

[312] See for example, Peter G. Tsouras, *Third Reich Victorious: Ten Dynamic Scenarios in Which Hitler Wins the War* (London, UK: Lionel Leventhal Limited, 2002); and Peter G. Tsouras, *Disaster at D-Day: The Germans Defeat the Allies*, June 1944 (London, UK: Greenhill Books, 2000).

[313] Anne Huff, ed., *Mapping Strategic Thought* (Chichester, UK: Wiley, 1990), 16.

[314] Gerard P Hodgkinson and others, Nicola J. Bown, A. John Maule, Keith W. Glaister, and Alan D. Pearman, "Breaking The Frame: An Analysis Of Strategic Cognition And Decision Making Under Uncertainty," *Strategic Management Journal*, 20 (1999): 979. Cited hereafter as Hodgkinson and others, "Strategic Cognition."

through the use of this particular cognitive mapping technique" and found cognitive mapping to be "an effective means of limiting the damage accruing from this bias."[315]

Because Hoffman's approach is generalized from weather forecast modeling—which has become increasingly accurate at certain scales—it would be reasonable that against Type 1 domains this may be a useful approach. But, what of Type 2 domains? Given the appeal and excitement that come with attacking such mysteries, innovative experimentation remains the surest way to learn about how causality mapping might contribute to positive advances in intelligence sensemaking about Type 2 domains. Given the conclusions reached by Hodgkinson, it appears likely that intelligence professionals making sense of type 2 domains might mitigate framing effects through such diagramming.

The Roles of Intuitive Thinking in Intelligence Sensemaking

Given these considerations, what are (or should be) the roles of skills-based intuitive and heuristic-based intuitive thinking in intelligence sensemaking? Many, if not most, intelligence professionals have had a "feeling" about an issue and what is going to happen. Sometimes those intuitions are correct, particularly if the requirement entails real-time observation and situational awareness. When it comes to anticipatory sensemaking, however, the authors suspect that intelligence professionals may fare no better than does the average citizen in predictive situations.[316]

There are a number of reasons for this, not the least of which has to do with the availability of evidence, or relevant information. A somewhat persistent myth about intelligence is that its professionals have access to all the information they need and that they can get any and all other necessary information. This not only simply is not true but is likely highly undesirable. While it is true that intelligence professionals must make their assessments based on incomplete, often faulty, and sometimes deceptive information, at least they can do so. Forcing them to try to make sense of *all* the relevant information relating to an issue would likely burden them sufficiently so as to preclude anything but the most general findings being issued—if anything

---

315 Hodgkinson and others, "Strategic Cognition," 977, 979. Framing problems differently has been empirically shown to dramatically change the decisions people make. See Amos Tversky and Daniel Kahneman, "The Framing of Decisions and the Psychology of Choice," *Science*, vol. 211, no. 4481 (10 January 1981), 453-458.

316 A near synonym is "prediction." However, the term implies a future that can be anticipated along with the establishment of the likelihood (probability) that it can be determined. Rather than such probability juggling, what people really do is to anticipate ranges of situations. This is the meaning attributed to "anticipatory sensemaking."

can be issued at all. Finally, as was discussed above in relation to figure 1 (see Chapter 3), complexity, ambiguity, and uncertainty increase as one moves from Descriptive to Estimative (or Anticipatory) Intelligence.[317]

It is true that a so-called "smoking gun" may exist from time to time. For example, the images of the Soviet SS-4 MRBMs in Cuba collected on 15 October 1962 left little doubt as to what they were—but such occurrences are rare. Further, intelligence professionals compete against other foreign intelligence organizations whose professionals may be as skilled at obfuscating what their factions, groups, or nations are doing as we are at making sense of what they are doing. Sometimes that other side is, in fact, better. And, like a closely matched sports event, the difference between valid and true sensemaking versus invalid or untrue sensemaking—or even no sensemaking at all—might be the result of luck.

Whether or not intelligence professionals do indeed have any better predictive skills than non-professionals should be a testable question; however, it remains one of many questions about intelligence sensemaking that remain open because of gaps in the empirical foundation for a psychological theory of performance of intelligence sensemaking.[318]

Many intelligence professionals work in one or more Type 2 domains where it is far from easy to come up with good operational definitions to identify experts. For example, how does one determine that an intelligence professional is an expert? In Type 1 domains, as discussed above, this can be clear. Among the defining features are years and breadth of experience, but also measurable performance, and status within social networks (i.e., the expert is the "go-to" person for dealing with specialized problems).[319]

However, as Kahneman and Klein note, such is not the case when "experts" must work outside their favored domains or within ones that are unstable and do not provide the cues and feedback necessary for calibration.[320] Since such Type 2 domains are ones in which the primary task goals involve the understanding and prediction of the activities of individuals or groups, accuracy and precision are elusive. Consequently, as has been noted, Type 2 domains are also characterized by tasks involving a lack of timely feedback and a paucity of robust decision aids. Further, if the object of study

---

[317] How much information is truly necessary is discussed in more detail below.

[318] Robert R. Hoffman, "Biased about Biases: The Theory of the Handicapped Mind in *The Psychology of Intelligence Analysis*," Panel Presentation for "Designing Support for Intelligence Analysts," S. Potter, Chair, in *Proceedings of the 48th Annual Meeting of the Human Factors and Ergonomics Society* (Santa Monica, CA: Human Factors and Ergonomics Society, 2005), 409.

[319] Robert R. Hoffman, "How Can Expertise be Defined?: Implications of Research from Cognitive Psychology," in Robin Williams, Wendy Faulkner and James Fleck, eds., *Exploring Expertise* (New York, NY: MacMillan, 1998), 85.

[320] Kahneman and Klein, "Intuitive Expertise," 521.

does not know what she or he will do, how can someone else predict it reliably? Therefore, is it any surprise that in such domains, intuition is limited in its usefulness? What are intelligence professionals to do?

It should be reiterated that although over-estimative errors in intelligence sensemaking are unacceptable, under-estimative errors are even less tolerated. It is better to have warned and been wrong than not to have warned and been wrong. False alarms are better than misses. A warning about an attempt by another individual to set off a bomb on a subway system—which subsequently does not occur—generates far less uproar (if any at all) than does a failure to warn of an individual who in fact plans to blow up an airliner, and through anticipatory sensemaking, being able to catch him preemptively. In the former case, the fact of the warning may even be viewed as the measure by which the attack was prevented. The would-be terrorist was scared off because his plans were publicized.

Unfortunately, this bias toward false alarms feeds into the blame game that demoralizes intelligence professionals. This creates a perception of an inevitability of failure perhaps captured by the water cooler joke, "How many analysts does it take to change a light bulb?" For which the answer is three: "One to not notice it is burned out, one to not change it, and one to take the blame."

Does More Information Improve Anticipation?

What about foraging for more information? Paul Slovic's 1973 research into the impact on the reliability of bookmakers as they brought more information into consideration noted that it did not increase their accuracy (figure 2, above).[321] Myers, in commenting on the Minnesota meta-study, notes further that providing clinicians with additional information from files or interviews actually reduced their accuracy.[322] The question becomes how much information is *enough*? A related question is, "how does one know that in advance?" It must be concluded that even with more information, Type 2 intuition remains of uncertain validity.

It is by no means obvious that simply throwing more information at a problem will make solving it any easier. For instance, to investigate the question of apparent under-utilization of information, Phelps and Shanteau (1978) used two methods to examine the strategies of livestock judges. In one situation, judges were presented with photos of female pigs (gilts) and asked

---

321 Slovic, "Behavioral Problems"; also Fischoff, "Condemned to Study the Past."

322 Myers, *Intuition*, 173-174. See W. M. Grove and others, "Clinical Versus Mechanical Prediction: A Meta-Analysis," *Psychological Assessment*, vol. 12 (2000): 19-30. Paul Slovic came to the same conclusions in 1973; see Slovic, "Behavioral Problems."

to rate each on breeding quality (their usual task). The results revealed the same sorts of limitations reported by other research, with each judge apparently relying on only a few of the 11 possible cues (e.g., weight, length, ham thickness, heaviness of bone structure, freeness of gait, etc.). In the second condition, judges were presented with verbal descriptions of the gilts (supposedly based on a telephone conversation) that listed the levels of the 11 attributes for each animal. In this case, analysis of the judgments of breeding quality revealed a subtle pattern: Judges used between nine and eleven cues, which tended to interact. Combined with results from a post-experimental interview, the judgments revealed an underlying strategy involving two waves of information integration: Measures are collapsed into intermediate judgments including size and meat quality; these judgments are then combined into an overall judgment.

The difference in results for the two tasks is striking. With the pictures of gilts, the relevant stimulus attributes were naturally correlated and perceptually chunked (e.g., tall gilts tend to be heavier and wider). Thus, even though a judge may have perceived all of the cues, only one significant cue might be needed explicitly to generate a given intermediate judgment. With the verbal descriptions, on the other hand, the cues were presented separately, the expert had to work through them explicitly, and the effects of cue intercorrelations revealed.

Benjamin Kleinmuntz obtained a similar result using the "Twenty Questions" game and a set of test cases to add structure to interviews with intern and resident neurologists.[323] Experience made a big difference in diagnostic accuracy, of course, but also in the number of questions asked about symptoms in each test case. In order to diagnose a case, the advanced experts asked fewer questions but they also spent less time pursuing incorrect hypotheses. Indeed, experts tended to ask about symptoms that yielded the greatest amount of diagnostic information relative to their hypothesis, reflecting "economy of effort."[324]

---

323 Benjamin Kleinmuntz, "The Processing of Clinical Information by Man and Machine," in Benjamin Kleinmuntz, ed., *Formal Representations of Human Judgment* (New York, NY: Wiley, 1968), 149-186.

324 The field of expert reasoning has a particularly rich literature. See, for example Sylvia Scribner, "Studying Working Intelligence," in Barbara Rogoff and Jean Lave, *Everyday Cognition: Its Development in Social Context* (Cambridge, MA: Harvard University Press, 1984), 9-40; and Sylvia Scribner, "Thinking in Action: Some Characteristics of Practical Thought," in Ethel Tobach, Rachel Joffe Falmagne, Mary Brown Parlee, Laura M. W. Martin, and Aggie Scribner Kapelman, eds., *Mind and Social Practice: Selected Writings of Sylvia Scribner* (Cambridge, UK: Cambridge University Press, 1997), 319-337; K. Anders Ericsson, Neil Charness, Robert R. Hoffman, and Paul J. Feltovich, eds., *Cambridge Handbook of Expertise and Expert Performance* (Cambridge, UK: Cambridge University Press, 2006); and Robert R. Hoffman and Laura Grace Militello, *Perspectives on Cognitive Task Analysis: Historical Origins and Modern Communities of Practice* (Boca Raton, FL: CRC Press/Taylor and Francis, 2008).

Future Vision: Red Brains, Blue Brains?

Given the discussion to this point, it is not enough to continue to ask whether expertise hinders judgment, or whether more information improves prediction. One might ask more productively about how we might improve the circumstances for sound reasoning. A partial answer may lie in the work of a group of Southern California political scientists whose recent neurocognitive research addresses how people behave and make predictions. For example, Darren Schreiber *et alia* used functional MRIs (magnetic resonance imaging tests) to assess how people associated with the U.S. Republican and Democratic political parties deal with risk. The researchers discovered that members of the two groups used distinctly different portions of their brains when making "winning risky versus winning safe decisions."[325] The authors note that the different portions of the brain play different roles in human cognition and conclude that

> it appears in our experiment that Republican participants, when making a risky choice, are predominantly externally oriented, reacting to the fear-related processes with a tangible potential external consequence. In comparison, risky decisions made by Democratic participants appear to be associated with monitoring how the selection of a risky response might feel internally.[326]

While neurocognitive paradigms for intelligence sensemaking have not yet formally been identified or established, implications of this work—to the degree that intelligence professionals can speak to the concerns of decisionmakers who are, after all, particular political partisans—may be significant. The research to date shows that the cognitive mechanisms and especially the emotion-based attitudes of partisan sensemakers shape their reasoning as they assess uncertain and risky phenomena.

Additional research could explore biologically based differences that correlate with the strong intuitive tradition by which sensemakers (and others) analyze, synthesize, and interpret intelligence phenomena. For example, in what situations are the skills and services of externally oriented intelligence professionals required and in what situations are those of internally oriented professionals needed? Similarly, in communicating intelligence results to

---

[325] Darren M. Schreiber and others, "Red Brain, Blue Brain: Evaluative Processes Differ in Democrats and Republicans," Paper delivered at the 2009 American Political Science Association Meeting, Toronto, CA, URL: <http://ssrn.com/abstract=1451867>, accessed 9 December 2009. Cited hereafter as Schreiber and others, "Red Brain, Blue Brain." While it is too soon to determine the impact of their (and similar) work, it may prove revolutionary to political affiliation studies.

[326] Schreiber and others, "Red Brain, Blue Brain."

consumers, are briefers more effective if their neurocognitive "risk" strategies match those of their consumers?[327]

Looking Ahead

Intuitive reasoning is something that we do naturally, all the time. It cannot be prevented, is not easily neutralized, and it is sometimes useful and necessary in sensemaking. While it can be reliable when employed as the sole basis for actionably valid, predictive intelligence creation in Type 1 domains, it is highly fallible when used for intelligence creation in Type 2 domains.

What can be done to challenge and validate the surety one has about an intuitive intelligence judgment? Employing approaches to reasoning such as those found in critical thinking seminars and courses, especially as currently offered across the IC's educational institutions, and developing skills that aid mindfulness (as discussed in the Preface), offer possible means of accomplishing calibrated reasoning.[328] Diagramming such as was done in figure 4 may challenge and certainly will augment the reasoning. But what else can be done to increase certainty that derived conclusions are valid? This becomes the subject of the next chapter.

---

[327] An important caveat is that just because one person exhibits a particular "style" of dealing with risk does not make them unsuited for other situations where a different approach may seem to be warranted. Such reasoning smacks of "biological determinism" and is no more appropriate in intelligence work than it was in the domains discussed in Richard Hernstein and Charles Murray's *The Bell Curve: Intelligence and Class Structure in American Life* (New York, NY: The Free Press, 1996). For a history and in-depth discussion of scientific determinism (sometimes also referred to as scientific racism) see Stephen J. Gould, *The Mismeasure of Man* (New York, W. W. Norton & Company, 1996).

[328] That these approaches will be effective admittedly is a hypothesis in need of testing in real-life situations. In experiments with students, Jennifer Reed (among others) finds that critical thinking does improve the quality of judgment and decision-making. The handy difference between the world in which intelligence professionals find themselves and Reed's classroom setting is that while there are no answers in foresight in the former, truth can be known in advance in the latter. See Jennifer H. Reed, *Effect of a Model For Critical Thinking on Student Achievement In Primary Source Document Analysis And Interpretation, Argumentative Reasoning, Critical Thinking Dispositions, And History Content in a Community College History Course*, PhD Dissertation, College of Education, University of South Florida, December 1998, URL: <http://www.criticalthinking.org/resources/JReed-Dissertation.pdf>, accessed 2 February 2010.

CHAPTER 6
Considering Validation

How does one know if the knowledge that intelligence sensemakers create is itself valid? Does accuracy alone ensure validity? What was accurate when findings were communicated may not be accurate subsequently. This flux suggests a strong procedural basis for validation. For example, were steps followed to avoid perceptual errors and cognitive traps? Was the process documented? Were alternatives adequately explored? Given the inherent uncertainty in intelligence judgments, it remains possible that all the appropriate processes may be sufficiently applied and yet the judgment is wrong—cold comfort for relatives of the victims and the survivors if the result is a terrorist attack on a scale of those in September 2001. By exploring validation perspectives from cognate fields, we may advance our parochial understanding.

Analogies from Other Fields

Medicine

Medical practice is at times presented as having notable similarities to intelligence practice. For example, with respect to validation, an ultimate metric for failure in medicine is that the patient dies. But is medicine successful if the patient lives? At what quality of life and for how long are two additional questions. Perhaps death with a minimum of suffering is the most favorable medical outcome—is this a success? Depending on the specifics of the case, maybe it is. Is there a difference if the patient is very old or very young? The author's own experience with the death of his mother is that elderly, somewhat frail patients—in practice—do not seem to receive the same level of treatment as do younger but equally severely ill patients. In the former case the patient, the author's mother, died within about 72 hours without what appears to be undue suffering but also without significant treatment of the condition, only an easing of the symptoms. In the latter case, the child recovers. Are both treatment regimes successful? Does the inevitable or perhaps the perceived inevitability of death by the attending physician become a factor in determining the treatment?

In medicine, success is measured post-facto. Early organ transplants were successes if the patients lived only for a brief period afterward. Other factors include quality of life, longevity, and costs. A recent T-cell-based, bioengineered trachea is considered a significant success not only because it succeeds but also because it dramatically improves the patient's quality of life:

> the successful outcome shows it is possible to produce a tissue-engineered airway with mechanical properties that permit normal breathing and which is free from the risks of rejection seen with conventional transplanted organs. The patient has not developed antibodies to her graft, despite not taking any immunosuppressive drugs. Lung function tests performed two months after the operation were all at the better end of the normal range for a young woman.[329]

If, a year later, however, the patient dies (from a related cause) was the expense worth it? Certainly, her family can be expected to indicate this is the case. But what of the cost to the larger society and perhaps other patients suffering from other ailments, who cannot get necessary resources because they are tied up in this specific treatment? Viewed in an intelligence context, if a terrorist attack is thwarted (characterized as a success) but a year later a larger and more devastating attack occurs, and it does so *because* the earlier attack was prevented, was the earlier disruption a success?

Jurisprudence

Jurisprudence is an adversarial system in which the ultimate confrontation is a trial wherein two advocates make inferences about evidence to argue opposite sides of a case before an impartial third entity or body (often of non-experts). This is the system at its best. In practice, the skill level of the advocates may vary. One may be more skilled or more proficient than the other. The impartial body (judge, jury) may be misled to a decision. Were this not the case then there would never (or very rarely) be cases where innocent people are convicted, imprisoned, and sometimes executed.[330]

While such failures appear to be rare, they may be examples of a limited consideration of evidence where only one side of an issue is examined. The fallacy is not new. Diagoras of Melos (5th Century BCE) was confronted with votive offerings carved by sailors in gratitude to the gods at their safe return from the sea. Diagoras unpopularly observed that only those who had

---

329 University of Bristol, "Adult Stem Cell Breakthrough: First Tissue-engineered Trachea Successfully Transplanted," *Science Daily*, 18 November 2008, URL: <http://www.sciencedaily.com/releases/2008/11/081119092939.htm>, accessed 21 November 2008.

330 For more on exonerations see Samuel R. Gross and others, "Exonerations in the United States, 1989 through 2003," *Journal of Criminal Law and Criminology* 95, no. 2 (2005): 523-560.

returned carved the votives; those who had not returned [the missing class of evidence] carved none.[331]

A success metric involving law would be one where either no challenge was made or it was repudiated and the convicted person was in fact guilty. But, what if the person is found innocent, or at least not guilty? They may be truly innocent. However, the prosecuting attorney may be incompetent, the evidence circumstantial or otherwise incomplete. In this case the actual innocence of the person may be independent of the findings of the court.

So, how does one measure success? There are at least five points of view involved in jurisprudence: That of each of the advocates, the judging entity (a jury or judge), the accused person, and the community or government. Each party, depending on the verdict, has a different metric for success. In certain types of cases such as those involving child molestation or alleged terrorism, the accused person tends to be deemed guilty by the community, prosecuting advocate, and government even if exonerated. In all cases where an opinion—particularly in the media—runs counter to the majority's views, the conclusion may be made that the court failed to render the "right" verdict as was also seen in the highly publicized cases of motorist Rodney King and the murder trial of O.J. Simpson.

Science

Science involves a number of metrics that include a sound method of documenting both process and results, as well as replication. Work is considered preliminary and non-definitive if it has not been replicated. As a typical example, in writing about the need for studies of the genetic bases of fidelity among humans, the editors of a collection of articles on genetics of behavior in *Science* refer to a comment by Australian psychologist, Simon Easteal, that despite the intellectual appeal of theories about genetic links to specific behaviors, "there are few replicated studies to give them heft."[332] In other words, the underlying theories may not be sound.

Science depends on refutation of alternative hypotheses, and replication studies attempt to refute that which has been shown. It is quite acceptable to be wrong so long as one admits it when the fact becomes apparent. Indeed, one model for science is that of competitive cooperation. Scientists attempt to tear down the new work of colleagues—without resorting to personal attacks. This dialectic approach may last generations or longer. In the process new knowledge is discovered and—if it cannot be refuted—validated.

---

331 Taleb, *The Black Swan*, 100.
332 Constance Holden, "Parsing the Genetics of Behavior," *Science*, vol. 322, no. 5903 (7 November 2008), 893.

Replication in Intelligence

The inability to replicate much of the process of sensemaking in intelligence limits the application of this indispensable practice of science. The pressures of real-time production inhibit the re-visitation of past judgments, although with at least one recent National Intelligence Estimate, the repetition of "alternative analysis" led to the questioning and revision of the original conclusions. Essentially, any meaningful replication of intelligence phenomena can only be accurately made in foresight, as in a National Intelligence Estimate. It is not replication if one group of practitioners performs sensemaking of an event before it happens and another group does so afterward.[333] Therefore, replication needs to occur *before* an event occurs.

When replication of methods against the same problem occurs it is typically in the intelligence school setting. It is the author's experience that such efforts yield a common set of explanations with some outliers.[334] For example, when intelligence students faced with a scenario involving three fictitious nations at odds with each other develop a common set of hypotheses regarding who will initiate a war and with whom, and are then given a finite set of evidence and a common method such as the Analysis of Competing Hypotheses, they come to similar conclusions as to which hypotheses are the least likely and therefore which eventualities can be expected.[335] Unfortunately there does not yet exist a similar body of results for real intelligence problems interpreted through the lenses of different intelligence disciplines and sources.

Yet, replication remains an important metric of the intelligence sensemaking process. As Caroline Park notes, "[the] basic reason research must be replicated is because the findings of a lone researcher might not be correct."[336] In the context of intelligence, un-reviewed and even reviewed conclusions of an intelligence professional may simply be incorrect. Admittedly a supervisory review process that grows more stringent as increasingly significant

---

[333] The problem is that latter group has the benefit of being aware of what actually happened, a factor they are unable to ignore in their deliberations. Since the two groups therefore are working in different contexts, replication has not occurred.

[334] The author observed this over a period of eight years of teaching new intelligence professionals at the National Security Agency in both a new employee orientation program and in the author's critical thinking and structured analysis course. At least 1,000 individuals have participated in the two courses the author offers. For more information on the course see Moore, *Critical Thinking*.

[335] It is the outliers that are truly fascinating in this classroom experience. Unfortunately, given the context imposed by a classroom setting and operational constraints, it has so far not been possible to capture why some students offer the outlying positions.

[336] Caroline L. Park. "What is the Value of Replicating other Studies?" *Research Evaluation* 13. No. 1 (December 2004), 198. Cited hereafter as Park, "Value."

implications emerge from the intelligence conclusions minimizes the likelihood of error. Still, errors do occur. One replicative method known within the IC, that of "Team A – Team B," appears not to be widely practiced although it has been used on specific issues.[337]

But what of the methods employed in the intelligence sensemaking process itself? Intelligence sensemaking can involve both quantitative and qualitative methods. As Park observes, quantitative research "can be replicated with great accuracy and precision."[338] Intelligence conclusions that result from counting observed phenomena such as aircraft located around an airfield can easily be repeated. But repeatedly and consistently measuring the intentions of the owners of those aircraft—the object of qualitative sensemaking—is more difficult although, as has been noted, not impossible.

Of note, however, is a danger of repetition confirming false results. Lynn Hasher, David Goldstein, and Thomas Toppino concluded that confidence in assertions increases through repetition of the assertions in situations when it is impossible to independently determine their truth or falsity.[339] Since intelligence evidence harbors a degree of uncertainty, repetition of evidence or even findings in proceedings designed to *confirm* their validity will only increase confidence that they are valid *irrespective of whether this is actually the case*. Ralph Hertwig, Gerd Gigerenzer, and Ulrich Hoffrage further show this "reiteration effect" is also part of evaluations made in hindsight.[340]

Validation in Foresight and Hindsight

People—and intelligence practitioners and their customers are merely people — evaluate judgments they have made in hindsight. They believe, according to Mark and Stephanie Pezzo, "that one could have more accurately predicted past events than is actually the case."[341] Thus, hindsight occurs at least in part because as people make sense of "surprising or negative" events,

---

337 Critical references to one effort commissioned by (then) DCI George Bush in 1976 regarding an NIE on Soviet Strategic Objectives are at URL: <http://intellit.muskingum.edu/analysis_folder/analysissov_folder/analysissovteams.html>, accessed 28 May 2010. In the referenced case CIA analysts conducted one analysis while a team of outside experts conducted an identical analysis.

338 Park, "Value," 190.

339 Lynn Hasher, David Goldstein, and Thomas Toppino, "Frequency and the Conference of Referential Validity," *Journal of Verbal Learning and Verbal Behavior*, vol. 16 (1977), 107. This work has been validated by Ralph Hertwig, Gerd Gigerenzer, and Ulrich Hoffrage, "The Reiteration Effect in Hindsight Bias," *Psychological Review*, vol. 104, no. 1 (1997): 194-202.

340 Hertwig, Gigerenzer, and Hoffrage, "The Reiteration Effect in Hindsight Bias," 194.

341 Mark V. Pezzo and Stephanie P. Pezzo, "Making Sense of Failure: A Motivated Model of Hindsight Bias," *Social Cognition*, vol. 25, no. 1 (2007), 147. Cited hereafter as Pezzo and Pezzo, "Making Sense of Failure."

"the reasons in favor of the outcome [are] strengthened, and reasons for alternative outcomes [are] weakened."[342] Further, in hindsight all the relevant facts may be known whereas in foresight this is not the case. But evaluating "mistakes" in hindsight obscures an important point made clear by Taleb: Mistakes can only be determined as such by what was known at the time they were made and then only by the person making the mistake.[343] In other words, mistakes need to be evaluated from the points of view held in foresight. And seen from that perspective they may not be mistakes at all.

Applied to intelligence sensemaking, this means that many so-called intelligence errors and failures may, in fact, be well-reasoned and reasonable judgments based on what is known prior to the decision. Certainly, when viewed in hindsight they were wrong. But in foresight they were accurate and valid to the best of the sensemaker's abilities. How can this enduring problem be mitigated? One means involves making the process of sensemaking as deliberate and thorough as possible.[344] Doing so may reduce mistakes and failures as more alternative possibilities are considered and assessed. However, achieving this objective requires that the underlying practice be valid. Finally, it assumes that key evidence is knowable and known.

Validating the Practice of Intelligence Sensemaking

What else contributes to bringing about validated sensemaking? If a method does not do what it is commonly purported to do, is it invalid? This is one question that has been raised with regard to the Analysis of Competing Hypotheses, or as it is commonly known, ACH. Richards Heuer, Jr. initially developed the method for the detection and mitigation of attempts at adversarial denial and deception.[345] ACH forces consideration of alternative explanations for, or predictions about, phenomena.[346] It forces consideration of the entire set of evidence, not "cherry-picked" trifles that support a favored hypothesis. A common belief is that ACH mitigates what is sometimes

---

342 Mark V. Pezzo and Stephanie P. Pezzo, "Making Sense of Failure," 148-149. Pezzo and Pezzo draw on work by David Wasserman, Richard O. Lempert, and Reid Hastie in "Hindsight and Causality," *Personality and Social Psychology Bulletin*, vol. 17, no. 1 (February 1991), 30-35.

343 Nassim Nicholas Taleb, *Fooled by Randomness: The Hidden Role of Chance in Life and in the Markets*, 2nd revised edition (New York, NY: Random House, Inc., 2004, 2008), 56.

344 One means of accomplishing this is discussed below in Chapter 8.

345 Heuer did not invent this method. The earliest use of multiple hypotheses in examining phenomena was by Thomas C. Chamberlin prior to 1890. See Thomas C. Chamberlin, "The Method of Multiple Working Hypotheses," *Science*, vol. 15 (old series), no. 366 (7 February 1890): 92-96. Another former CIA employee, Morgan Jones, offers a version of ACH he calls hypothesis testing. See Morgan D. Jones, *The Thinker's Toolkit: 14 Powerful Techniques for Problem Solving*, revised edition (New York, NY: Crown Publishing, 1998). Jones' book also introduces the concept of and phrase "structured analysis."

346 Heuer, *Psychology*, Chapter 8.

known as the "Confirmation Bias," whereby people seek to prove a favored hypothesis through (among other things) selective exposure and selective perception. Allegedly it does so by asking people to think in a disconfirmatory fashion. They are to use the available evidence to disprove as many of the existing hypotheses as possible. However, a study by MITRE failed to show that it does eliminate the confirmation bias.[347] Both the MITRE study and an earlier one by NDIC student Robert Folker do suggest that ACH is of value when used by novice intelligence professionals. However, Folker tentatively concluded that experts seem not to be aided by the method.[348] Is it still a valid method for intelligence sensemaking?

Perhaps it is. The method provokes detailed consideration of the issue and the associated evidence through the generation of alternative explanations or predictions and the marshaling of the evidence. It asks the sensemaker to establish the diagnosticity of each piece of evidence. In some versions of the method, evidence is weighted based on source and relevance. Some computer-assisted versions of ACH under review for application in the IC consider the likelihood that the sensemaker has omitted a relevant—and perhaps the correct—explanation. Therefore, it can prompt a broader yet more detailed *sensemaking* of the issue than might otherwise occur.

ACH further makes explicit the fact that evidence may be consistent with more than one hypothesis. Since the most likely hypothesis is deemed to be the one with the least evidence against it, honest consideration may reveal that an alternative explanation is as likely or even more likely than that which is favored. The *synthesis* of the evidence and the subsequent *interpretations* in light of the multiple hypotheses is also more thorough than when no such formalized method is employed.

Indeed, Robert Folker's "modest" experiment in applying qualitative structured methods—specifically ACH—to intelligence issues showed that "analysts who apply a structured method—hypothesis testing, in this case—to an intelligence problem, outperform those who rely on "analysis-as-art," or the intuitive approach."[349] Simply put, Folker experimentally showed that method improves the quality of practitioners' findings. Folker's study offers evidence that "intelligence value may be added to information by investing some pointed time and effort in analysis, rather than expecting such value to arise as a by-product of 'normal' office activity."[350]

---

347 Cheikes and others, *Confirmation Bias*, iii.

348 MSgt Robert D. Folker, Jr., USAF, *Intelligence Analysis in Theater Joint Intelligence Centers: An Experiment in Applying Structured Methods*, Occasional Paper Number Seven (Washington, DC: Joint Military Intelligence College, 2000). Cited hereafter as Folker, *Experiment*.

349 Folker, *Experiment*, 2.

350 Folker, *Experiment*, 2.

One variation on ACH implementation provides a structured means of developing issues. As applied by the faculty and students of the Institute for Intelligence Studies at Mercyhurst College, practitioners begin with a high-level question and use sequential iterations of ACH to eliminate alternative explanations.[351] The next phase takes the "non-losers" and develops them further. Another round is conducted and again the "non-losers" are selected and further developed. While this could generate a plethora of branching explanations, in reality it is the author's experience that it tends to disambiguate the issue fairly efficiently. At worst the structuring inherent in the method leaves the sensemaker with an in-depth understanding of the issue; at best, a couple of eventualities and their likely indicators are determined. Assets can then be tasked, foraging conducted, and more exact determinations made as the issue develops.

The question remains: Is the method valid? This question generates considerable discussion among practitioners. When used honestly, the method certainly prompts a more thorough assessment of the issue.[352] Evidence is considered singularly and severally. Alternative explanations for the phenomena are taken seriously. New possibilities are discovered. Even the skeptical MITRE team noted some validity in the method, whereby

> participant assessment of new evidence was significantly impacted by beliefs they held at the time evidence was received. Evidence confirming current beliefs was given more "weight" than disconfirming evidence. However, current beliefs did not influence the assessment of whether an evidence item was confirming or disconfirming.[353]

Appropriateness, flexibility, and ease of use are other criteria that need to be established with respect to the various sensemaking methods. Whether the method facilitates foraging, analysis, synthesis, interpretation or some combination of the foraging and understanding processes, some means is needed to operationalize validation procedures so that intelligence practitioners can increase their efficiency and accuracy.

Since Congress (in 2004) directed the Intelligence Community to employ "Alternative Analysis" in its intelligence deliberations, structured analytic methods now are taught to all new intelligence sensemakers.[354]

---

351 Kris Wheaton and others, *Structured Analysis of Competing Hypotheses* (Erie, PA: Mercyhurst College, 2005).

352 The issue of "honest use" is nontrivial. Less-than-scrupulous sensemakers are certainly free to cherry-pick favorite evidence and go through the motions of considering alternative hypotheses. Then, once they have eliminated competing alternatives, their favorite remains. The difficulty for them is what to do with the audit trail the method produces.

353 Cheikes and others, *Confirmation Bias*, iii.

354 U.S. Congress, IRTPA, 2004, 33. The phrase "Alternative Analysis" is interpreted as meaning structured analytic methods.

Many more experienced professionals also receive education and training in such methods.

Johnston observed that the IC has at its disposal "at least 160" [analytic methods]…but it lacks "a standardized analytic doctrine. That is, there is no body of research across the Intelligence Community asserting that method X is the most effective method for solving case one and that method Y is the most effective method for solving case two."[355] As referenced here, such a doctrine arises out of knowledge that the specific methods are valid, in other words it has been demonstrated *empirically* that they actually do what they claim to do. Such a doctrine proffers a menu of sensemaking options dependent on the goals of the sensemaker. The current model, where validity is presumed by intelligence professionals because they are taught the method(s) in the community's training schools, is insufficient because—at the most basic level such a metric is insufficient for determining validity. Further, there is little sense of what methods are appropriate in what situations. A claim of "we always do it that way," is known to be insufficient but remains part of the "corporate analytic tradecraft."

Heuer noted that "intelligence [error] and failures must be expected."[356] One implication of this assertion is that intelligence leadership cannot fall back on a "lack of skills" excuse when the next major intelligence failure occurs. However, without validating a canon of method and a taxonomy to characterize its use, intelligence professionals will remain hamstrung in their efforts to make fuller sense of threatening phenomena, increasing the likelihood of error and failure. It is reasonable to presume Congress will not let the community commit failures similar to those of the past eleven years without severe repercussions.[357]

Seeking Validation: Toward Multiple Methods

Within the canon of social science method lies an approach to sensemaking that may offer intelligence practitioners a means of disambiguating the wicked mysteries as well as the hard puzzles they face daily. Even in current practice, intelligence practitioners employ this approach when they do not rely on merely one method for sensemaking. Multi-method intelligence

---

[355] Johnston, *Analytic Culture*, xviii.

[356] Heuer, *Psychology*, 184.

[357] As has been noted, such an *intelligence* failure occurred on 25 December 2009 in the skies over Detroit, Michigan (see URL: <http://intelligence.senate.gov/100518/1225report.pdf>). Two days after the release of the Senate report, 20 May 2010, partly as a result, Director of National Intelligence Dennis Blair left office at the request of the President. Even though the attempted attack was thwarted, it is reasonable to expect that intelligence professionals will be required to consider more possibilities and examine more information in the hopes they will notice its evidentiary value. They will need to be more imaginative.

sensemaking explores complex issues from multiple perspectives. Each method used—such as ACH—provides an incomplete understanding of the issue, leaving the intelligence professional the task of making sense of the differing sensemaking conclusions. While the results of different methods may converge, reinforcing a particular understanding of a phenomenon, they may also diverge and yield different interpretations. It is up to the intelligence professional to resolve and make sense of the differences.

For example, intelligence professionals who engage in a "multi-frame" sensemaking approach consider issues from multiple points of view created from the intersections of action- and process-focused vantage points and the perspectives of the individual and the collective. As developed by Monitor 360 for the National Security Agency's Institute for Analysis, it facilitates sensemakers' developing different answers to the intelligence question at hand.[358] They must combine the differing results, in other words, synthesize and interpret partial answers, in order to better understand the issue underlying the question and to determine a best (at the time) understanding of the issue.

The lexicon of multi-methodology provides a term for this combinatorial activity: triangulation, or pinpointing "the values of a phenomenon more accurately by sighting in on it from [the] different methodological viewpoints employed."[359] This is a process of measurement—which to be useful (accurate) "must give both consistent results and measure the phenomenon it purports to measure."[360] In other words, triangulation requires that the methods employed are repeatable and valid. Intelligence creation requires that those methods be applied with rigor.

Fortuitous circumstances allow the authors to present in the next chapter a multi-method case study of sensemaking in a Type 2 environment that illustrates the interplay of intuition, logic, analysis, synthesis and interpretation of an issue of interest to the National Intelligence Council.

---

[358] While the approach is unclassified, it was first presented at an NSA (CLASSIFIED) conference on analysis. The author subsequently worked with a team from Monitor 360 to create a training course through which intelligence professionals could learn and apply this approach.

[359] John Brewer and Albert Hunter, *Foundations of Multimethod Research: Synthesizing Styles* (Thousand Oaks, CA: Sage Publications, 2006), 5. Cited hereafter as Brewer and Hunter, *Multimethod Research*.

[360] Brewer and Hunter, *Multimethod Research*, 5.

CHAPTER 7
Making Sense of Non-State Actors: A Multimethod Case Study of a Wicked Problem[361]

David T. Moore, with Elizabeth J. Moore, William N. Reynolds, and Marta S. Weber

> *A foolish consistency is the hobgoblin of little minds, adored by little statesmen and philosophers and divines. With consistency a great soul has simply nothing to do. He may as well concern himself with his shadow on the wall.*
>
> — Emerson, *Essay on Self-Reliance*

Introduction

In the pragmatic U.S. tradition invoked by Emerson, we question the utility of consistently following the Sherman Kent model where one or many solitary scholars try to work out the solution to problems. We also recognize that team-based intelligence production suffers from inherent drawbacks.[362] However revolutionary it may be, we find that a diversely practiced, multimethod approach that does incorporate a specific process, organizing principles, and an operational structure can fulfill the need for 21st century intelligence sensemaking. Such an approach reflects a Kendallian approach to intelligence sensemaking: It collaboratively paints a picture for a decision-maker rather than presenting a "scientific fact."

---

[361] The authors would like to thank the following individuals for their assistance in the research that led to this chapter: David Colander, Richards J. Heuer, Jr., Robert K. Hitchcock, James Holden-Rhodes, Donald McGregor, Suzanne Sluizer, Joseph Tainter, Kristan J. Wheaton, and students in the 2007-2008 cohort of the Mercyhurst College Institute for Intelligence Studies. Thanks are also due to the participants of the Eurasia Group-sponsored conferences for their insights.

Work done by Least Squares Software was funded by IARPA under AFRL Contract Number FA8750-07-C-0312.

[362] Johnston, *Analytic Culture*, 68-70. Johnston finds that "without specific processes, organizing principles, and operational structures, interdisciplinary teams will quickly revert to being simply a room full of experts who ultimately drift back to their previous work patterns."

The case study of sensemaking presented here is not merely a thought-experiment. It rests on a shortfall in understanding the security implications of non-state actors. In this instance, an original National Intelligence Council study prompted a further examination by an academic intelligence studies department and a government contractor. In all three studies partial sense of the issue was made. As we explore their independent work, we seek to apply intelligence rigor and discuss—admittedly in hindsight—how separate parts of the problem could benefit from a triangulation of concepts and approaches to provide a coherent, larger view of the security role of non-state actors.

Introducing the Wicked Problem of Non-State Actors

It is commonly thought that non-state actors are emerging as a dominant global force in the realm of national and international security, yet conclusive evidence confirming this belief is lacking. Part of the challenge is that non-state actors fit the profile of a wicked problem. While it is true they can be identified (the name accomplishes this—they are *non-state* versus *state* actors), there is no commonly accepted definition.[363] In other words, non-state actors are defined by what they are not, leaving room for disagreement as to what they are. Further, some non-state actors can be characterized as "good" and others as "bad." Differing points of view about whether a particular non-state actor is "good" or "bad" leads to varying characterizations of their activities. Attempting to "solve" a non-state actor problem leads to good or bad solutions that may provoke unanticipated (and undesired) consequences. For example, U.S. attempts to eliminate threats posed by Al Qaeda gave rise to a different and unexpected problem, that of a globally distributed network of Al Qaeda and Al Qaeda "wannabes." An argument also could be made that the apparently unique globalization of a terrorist organization was also unpredictable in foresight.[364]

---

[363] National Intelligence Council, "Nonstate Actors: Impact on International Relations and Implications for the United States," Conference Report, August 2007, URL: <http://www.dni.gov/nic/confreports_nonstate_actors.html>, accessed 10 May 2010, The Conference Report suggested that "**Nonstate actors** are non-sovereign entities that exercise significant economic, political, or social power and influence at a national, and in some cases international, level. There is no consensus on the members of this category, and some definitions include trade unions, community organizations, religious institutions, ethnic groupings, and universities in addition to the players outlined above" (p. 2, emphasis in original)

[364] In hindsight it seems obvious that, in an era when globalization is a driving force, a terrorist organization would naturally become a globalized phenomenon. However, we must adopt the point of view we enjoyed prior to 11 September 2001. At that time what occurred was largely unlooked for and unpredicted. Most of us simply could not conceive of such a group as being globalized or taking advantage of globalized resources. Therefore, in that context, it was unpredictable and we were surprised.

Issues involving non-state actors lack clear definitions and are resistant to traditional intelligence approaches due to their open-ended nature; potential solutions to problems are neither clearly right or wrong; and difficult-to-discern and complex inter-linkages exist, although drivers for issues involving non-state actors can be identified (see figure 6).

Once an insurgency or terrorist campaign (or any other non-state actor activity) begins, the issue shifts from being a merely wicked problem to a combination of wicked and tame problems. Some aspects of the issue remain wicked—ill-defined, no right or wrong solution, open-ended, and so on. However, other aspects of the issue are tame, although difficult to make sense of. For example, the tactics likely to be used in an insurgency are finite and understandable and making sense of them is a tame, bounded process. Yet countering them invokes a series of new wicked problems encapsulated in the larger issue and likely to lead to unanticipated consequences (many of them wicked problems in their own right)—as the United States discovered with its post-9/11 dealings with Al Qaeda.

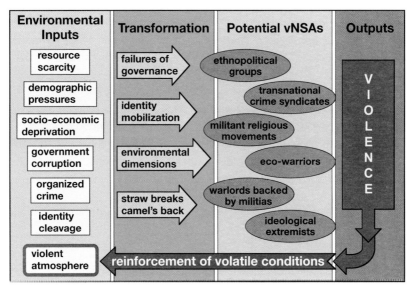

Figure 6. Drivers for the Rise and Growth of Violent Non-State Actors (vNSAs).
Sources: William Reynolds *et alia*, "Social Science Modeling Workshop: Understanding Iraqi Non-State Actors," Workshop Proceedings, Least Squares Software, Albuquerque, NM, 15 February 2008 (proprietary, used with permission). Image derived from Troy S. Thomas, Stephen D. Kiser, and William D. Casebeer, *Warlords Rising: Confronting Violent Non-State Actors* (Lanham, MD: Lexington Books, 2005).

Three Approaches to Making Sense of Non-State Actors

The starting point for this case study was a 2007 National Intelligence Council (NIC) Desktop Memorandum that analyzed key findings from a series of seminars co-hosted with the Eurasia Group, a global political risk research and consulting firm.[365] The Memorandum observes that non-state actors are of interest "because they have international clout, but are often overlooked in geopolitical analysis." The implicit but demanding questions of why and how much non-state actor "power" and "influence" have increased worldwide was not answered.[366] In light of this limitation, two follow on projects were undertaken to complement the NIC study. First, the Mercyhurst College Institute for Intelligence Studies examined the impact of non-state actors in Africa. Then, a social science workshop convened by Least Squares Software of Albuquerque, New Mexico, under the auspices of IARPA, considered whether the influence and impact of non-state actors on international relations can actually be measured.

Key Findings of the NIC Study on Non-State Actors

A series of NIC-Eurasia Group seminars in 2006 and 2007 discussed in exploratory fashion how the proliferation of non-state actors since the end of the Cold War "is transforming international relations."[367] As a collaborative, prospective assessment, the study found that

a globalization-fueled diffusion of finance and technology has enabled non-state actors to encroach upon functions traditionally performed by nation-states. This has facilitated their evolution into a

---

365 See the Eurasia Group web site, URL: <http://www.eurasiagroup.net/about-eurasia-group>, accessed 14 May 2010.

366 National Intelligence Council, "Nonstate Actors: Impact on International Relations and Implications for the United States," Conference Report, August 2007, URL: <http://www.dni.gov/nic/confreports_nonstate_actors.html>, accessed 27 April 2010, 2. Cited hereafter as NIC, "Nonstate Actors."

367 NIC, "Nonstate Actors," 1. The NIC study focused on so-called benign non-state actors—non-governmental organizations, multinational corporations, and super-empowered individuals—although it was impossible to have the discussions without frequent reference to terrorists, warlords, and other "malign" actors. The word "benign" is relative, but in this case refers to entities that at least give a nod to national and international institutions, laws, and norms. Nearly a decade earlier political scientists Philip Schrodt and Deborah Gerner derived similar conclusions about the post-Cold War explosion of *violent* non-state actors as agents of complex humanitarian crises. See Philip A. Schrodt and Deborah J. Gerner, "The Impact of Early Warning on Institutional Responses to Complex Humanitarian Crises," Paper presented at the Third Pan-European International Relations Conference and Joint Meeting with the International Studies Association, Vienna, Austria, 16-19 September 1998. Cited hereafter as Schrodt and Gerner, "CHC."

form unheard of even a few years ago. For example, "philanthrocapitalist" charities such as the Gates Foundation have greatly expanded notions of what a charitable NGO should look like.[368]

The discussions found that few non-state actors are completely independent of nation-states, and they do not have uniform freedom of movement. Further, non-state actors have the most latitude in either weak, or, at the other end of the spectrum, post-industrial states. The bulk of the world's population, however, lives in so-called "modernizing" states such as China, India, and Russia.[369] They remain entrenched in the class state system: firmly sovereign, centralized, and bureaucratic; using nationalism (including suppression of minorities) as an instrument of state power; and defining national security in terms of force. These nations have been highly effective in suppressing non-state actors or co-opting them through deployment of state-controlled substitutes including a subset of state-owned multinational corporations and enterprises—sometimes referred to as the mind-bending GONGO (government-operated non-governmental organization). The NIC-Eurasia Group discussions determined that the entity—aside from terrorists and criminals—that is most problematic for the United States is not technically non-state at all. It is instead the state-owned enterprise, often a front for advancing the interests of a modernizing-state government.

Finally, the NIC-Eurasia Group effort posited that the significance of "benign" non-state actors was that they propagated Western values in regions where these were absent. In this case, the problem was not that such actors were too powerful. Rather the opposite was the case: "in many parts of the world [benign non-state actors'] influence is limited—a factor that is contributing to the tilting of the global playing field away from the United States and its developed-world allies."[370]

Key Findings of the Mercyhurst Study on Non-State Actors

Students in the Mercyhurst College Institute of Intelligence Studies (MCIIS) focused on the roles non-state actors play and their expected impact in Sub-Saharan Africa over the next five years (*results*), and on building a multi-methodological paradigm for considering the issue (*process*).[371] Within this context, three additional questions were raised:

---

368 NIC, "Nonstate Actors," 1.
369 NIC, "Nonstate Actors," 1.
370 NIC, "Nonstate Actors," 1.
371 Mercyhurst College Institute for Intelligence Studies, "Terms of Reference: The Role of Non-State Actors in Sub Saharan Africa," Wikispaces.com, URL: < https://nonstateactorsafrica. wikispaces.com/Terms+of+Reference>, accessed 28 April 2010. Cited hereafter as MCIIS, "Terms of Reference."

- What is the likely importance of [Non-State Actors] vs. State Actors, Supra-State Actors and other relevant categories of actors in Sub-Saharan Africa?
- What are the roles of these actors in key countries, such as Niger?
- Are there geographic, cultural, economic or other patterns of activity along which the roles of these actors are either very different or strikingly similar?[372]

With respect to the first question, their research found that Africa can be organized into three geographical regions based on the roles that non-state actors play (figure 7): In Western Sub-Saharan Africa there are no clear trends in the roles played by non-state actors; in Central Sub-Saharan Africa, anti-government non-state actors are most active and likely to remain so over the next five years; and in Southern and Eastern Sub-Saharan Africa government

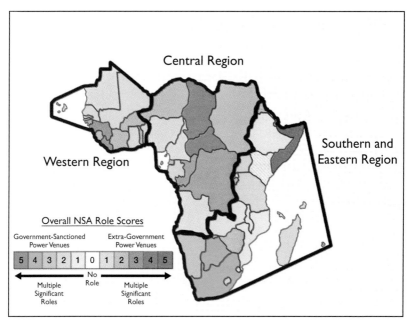

Figure 7. Composite Non-State Actor Role Scores for Africa.
Source: Mercyhurst College Institute for Intelligence Studies, URL: <https://nonstateactorsafrica.wikispaces.com/Are+there+geographic%2C+cultural%2C+economic+or+other+patterns+of+activity+along+which+the+roles+of+these+actors+are+either+very+different+or+strikingly+similar%3F>, accessed 2 March 2008.

---

[372] MCIIS, "Terms of Reference."

sanctioned non-state actors are likely to wield the most influence.[373] Additionally, there appeared to be a strong correlation between the number of government-sanctioned multinational corporations and the number of NGOs. Finally, it seemed that terrorist organizations in Africa preferred operating in "countries with relatively more than less state control."[374]

The students developed a scoring system for both lawful and unlawful non-state actors, in terms of the socio-political environment, and applied this index to all 42 Sub-Saharan African countries (figure 7). The scoring characterized the roles of non-state actors vis-à-vis government and non-government interactions based on four drivers: An "ease of doing business" variable and a contrasting "corruption perception" variable; a democracy variable and a contrasting failed states variable.[375] Stable and failing states were revealed to have differing interactions with non-state actors. In the former, non-state actors were lawful actors who tended to have government-sanctioned role potentials, whereas in the latter they were typically unlawful actors engaged in anti-government roles. Botswana (a stable state) and the Central African Republic (a failed state) were representative of each (shown in figure 8). Of greater interest were the indicators for Kenya (figure 9), because they signaled or *anticipated* stability issues. The assessment was born out by the events that occurred during and after the early 2008 elections.[376]

Mapping significant multinational corporations, NGOs, and terrorist organizations to specific countries as representative of non-state actor activity revealed correlations between role potential spectra and geospatial data, whereby each generally supported the other.[377] Thus, geospatial sense-making tended to confirm the conclusions derived from the non-state actor role spectra.

---

373 Mercyhurst College Institute for Intelligence Studies, URL: <http://nonstateactorsafrica. wikispaces.com/Key+Findings>, accessed 2 March 2008. Cited hereafter as MCIIS, "Patterns of Activity."

374 MCIIS, "Patterns of Activity."

375 MCIIS, "Process and Methodology." The roles of non-state actors were tracked on a scale anchored by scores from six sample totalitarian states: North Korea, Cuba, Iran, Syria, Myanmar, Laos, and Libya.

376 Kristan J. Wheaton, email to the author, 4 March 2008. Such foresightful activity by students is not unknown. Schrodt and Gerner report a similar result involving student predictions about the state of the Iraq-Kuwait crisis in December 1990. Kahneman and Klein's remark about some individuals being able to discern correctly patterns that others miss (Chapter 5) is probably relevant in explaining both phenomena. See also Schrodt and Gerner, "CHC," 5.

377 Mercyhurst College Institute for Intelligence Studies, "Non-State Actors in Sub-Saharan Africa 2007-2012 Outlook," URL: <https://nonstateactorsafrica.wikispaces.com/ Key+Findings>, accessed 2 march 2008. Cited hereafter as MCIIS, "Outlook."

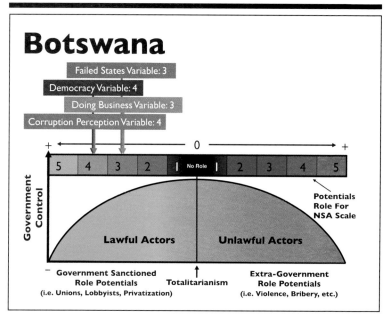

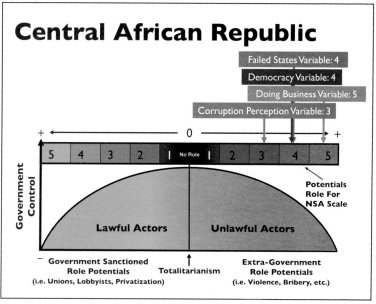

Figure 8. Non-State Actor Potential Spectra for Botswana and the Central African Republic.

Source: Mercyhurst College Institute for Intelligence Studies, "Role+Scores. ppt", URL: <https://nonstateactorsafrica.wikispaces.com/Model+for+SFAR>, accessed 2 March 2008. Cited hereafter as MCIIS, "Role Scores."

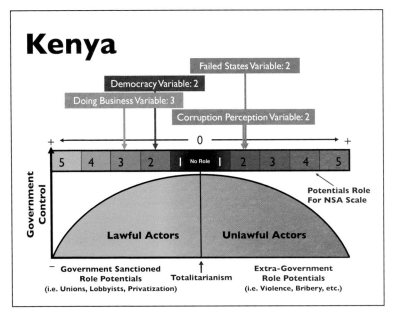

Figure 9. Kenyan Non-State Actor Potential Spectra.
Source: MCIIS, "Role Scores."

Key Findings of the Least Squares Study on Non-State Actors

The Least Squares study of non-state actors began with the hypothesis that "non-state actors emerge in vacuums and voids."[378] Their study focused on the issue of violent and non-violent non-state actors but also explored a set of contingent methodological approaches. The inquiry sought to contribute novel understanding of non-state actors by

synthesizing available data and disparate taxonomies,…by generating and testing hypotheses concerning the key dynamics driving the transfer of power from states to [non-state actors] and favoring the emergence of novel [non-state actors] under globalization; and…by investigating the development of methodologies that might be most useful for future research.[379]

Two key findings revealed the critical role of environmental knowledge and of public expectations in motivating non-state actors, both as individuals and as members of the collective. Such findings were found to be

---

378 Marta S. Weber, William N. Reynolds, James Holden-Rhodes, and Elizabeth J. Moore, *Non-State Actors in the Post-Westphalian World Order: A Preliminary LSS Inquiry,* Final Report for Air Force Research Laboratory (AFRL) Contract FA8750-07-C-031, March 2007, 6. Cited hereafter as Weber and others, *Non-State Actors.*

379 Weber and others, Non-State Actors, 6.

significant to efforts aimed at mitigating the recruitment of specific Al Qaeda-associated individuals to assail the United States. Additionally, the team found that an approach based on critical thinking led to reasoning pathways that likely would not have been noticed or explored had a less rigorous approach been employed.[380]

Approaches and Methodologies

Thinking Critically about the Issue

In order to impose structured thinking on a highly unstructured problem, the NIC advisor to, and the members of LSS first inventoried their own understanding of the non-state actor issue using eight of the elements of reasoning developed and espoused by the Foundation for Critical Thinking and used throughout much of the IC.[381] These elements include:

- Question at issue (What is the issue at hand?)
- Purpose of thinking (Why examine the issue?)
- Points of view (What other perspectives need consideration?)
- Assumptions (What presuppositions are being taken for granted?)
- Implications and consequences (What might happen? What does happen?)
- Evidence (What relevant data, information, or experiences are needed for assessment?)
- Inferences and interpretations (What can be inferred from the evidence?)
- Concepts (What theories, definitions, axioms, laws, principles, or models underlie the issue?)

A summary of several of the NIC and LSS perspectives on non-state actors based on this exercise are provided in table 4. Both groups agreed that a critical thinking approach was useful in developing a common understanding of the problem as it helped to ensure that the participants questioned their own thinking about non-state actors, rather than relying on previously held presumptions.

---

[380] These points are derived from William Reynolds' observations of and discussions with fellow LSS Workshop participants.

[381] Moore, *Critical Thinking*, 8-9; Richard Paul and Linda Elder, *The Miniature Guide to Critical Thinking: Concepts and Tools*, Sixth Edition (Dillon Beach, CA: Foundation for Critical Thinking, 2009), 3-6. Some practitioners in the U.S. Intelligence Community (among them one of the authors) in late 2008 expanded these eight elements of reasoning to ten based on the work of Gerald Nosich. Included were "alternatives," which makes explicit the fact that there are other ways to view an issue; and "context," which considers the fact that most issues or problems are themselves a piece of some larger issue or problem: There is a broader background for the issue that must be explored. See Nosich, *Learning to Think Things Through*, 96-98.

Table 4. Comparing NIC and LSS Critical Thinking Perspectives: Purpose, Points of View, and Assumptions

Purpose

NIC Perspective	*LSS Perspective*
To test whether the influence and impact of non-state actors on international relations can be measured; if so, whether they are serious competitors for power with nation-states; and the implications for US foreign policy.	ID role of non-state actors; Determine influence of non-state actors compared to state actors; Assess utility of sensemaking methodology/Multimethodology (MM) mechanisms of non-state actors role/influence; Develop/Define the concepts of non-state actors Influence/Impact; ID MM's useful for assessing impact/role of non-state actors; ID MM's/problem motifs of general interest to other research contracts;

Points of View

NIC Perspective	*LSS Perspective*

Although non-state actors seem to be more powerful than ever before in history, few of them act in total independence from nation-states and their influence and impact are highly dependent upon which part of the world is under discussion.
 	
Another (and common) point of view is that powerful non-state actors are universal and a serious threat to nation-states.
	"Victims"/Beneficiaries of non-state actor impact:
	Policy Makers
	   Minimizing Risk
	   Achieving Agenda
	Other non-state actors
	   Analyst—understand/advise
	   Methodologist/Researcher—Identify techniques for understanding
	      Useful methods
	      Understand/advise
	Sub interest structures within non-state actors and States

Table 4. Comparing NIC and LSS Critical Thinking Perspectives: Purpose, Points of View, and Assumptions (Continued)	
Assumptions	
NIC Perspective	*LSS Perspective*
Our point of view: implications for the US of the rise of powerful non-state actors depend upon which part of the world is under discussion. At times, our main concern may be that benign non-state actors are excluded from, or usurped by, certain states. The common point of view: all non-state actors in all parts of the world have serious implications for national governments (including that of the US).	Relation between nature of country and nature of economy—concept of *industrial* as an important discriminator; That we can induce truths from examples — commonalities -> analogy -> truth; Normative idea that pursuit of self interest is a driver in observed outcomes; The frame of a value system—suicide bombing makes different sense in different frames; Idea of a pro/con/fix or ACH type approach to value outcomes in different frames; *International* means "between nations." Are we only concerned with "international" phenomena?

Source: Participant notes, edited by the authors.

Literature Consultation

Concurrent with their critical thinking, MCIIS, LSS, and others examined key academic and applied-academic works related to the assessment of non-state actors. Notable among them, work by Bas Arts and Piet Verschuren describes a qualitative method for assessing the influence of stakeholders in political decision-making.[382] The "triangulation" referred to in their title encompasses "(1) political players' own perception of their influence; (2) other players' perceptions of the influence brought to bear; and (3) a process analysis by the researcher."[383] Arts and Verschuren tested this method through an assessment of the influence of NGOs on the 1992 Climate Convention. Their

---

[382] Bas Arts and Piet Verschuren, "Assessing Political Influence in Complex Decision-Making: An Instrument Based on Triangulation," *International Political Science Review*, vol. 20, no. 4 (1999): 411-424. Cited hereafter as Arts and Verschuren, "Assessing Political Influence."

[383] Arts and Verschuren, "Assessing Political Influence," 411.

work remains significant because attempts to measure or associate numbers with non-state actor power or influence have been so rare.

Another contribution in the applied realm came from recent work by a new generation of military (and ex-military) authors who see the rise of non-state actors as a seminal event that will drive U.S. national security strategy. Among these sources is *Warlords Rising: Confronting Violent Non-State Actors*, whose authors anchor their work in open systems theory (the concept that actors and organizations are strongly influenced by their environment).[384] In particular, they ask what environments give rise to violent[385] non-state actors, what sustains them, and how changes to those environments might disrupt them.

Application: Indicators of Non-State Actor Power in Africa

The project afforded the Mercyhurst team an opportunity to develop a promising new method for intelligence sensemaking and to catalogue its advantages and disadvantages (table 5). The students were able to validate their findings employing three different methods as well as different evidence sets and also assess their methodological validity. This kind of meta-sensemaking could constitute a bridge between now-traditional IC efforts and a revolutionary approach to building a sensemaking argument in official circles.

Of note is a remark by project supervisor Professor Wheaton: "The big advantage [of the multimethodological approach] was the ability to see similar patterns crop up again and again by looking at the data in different ways. This increased their [the students'] confidence enormously."[386] Additionally, given the temporal context (short) and the scale of the project (large) a multimethodological approach was perhaps the only means of tackling the problem. Finally, as Wheaton also notes,

> we may well be wrong about Liberia or some other country in the current study but we are unlikely to be wrong about every country and highly unlikely to be wrong about every country in the same direction. Assuming the model is reasonably accurate and given

---

384 Troy S. Thomas, Stephen D. Kiser, and William D. Casebeer, *Warlords Rising: Confronting Violent Non-State Actors* (Lanham, MD: Lexington Books, 2005). Other key books in this genre are General Rupert Smith, *The Utility of Force: The Art of War in the Modern World* (New York, NY: Alfred A. Knopf, 2007) and John Robb, *Brave New War: The Next Stage of Terrorism and the End of Globalization* (New York, NY: John A. Wiley and Sons, 2007).

385 *Warlords Rising* is focused on violent non-state actors, but the present authors find that the question of environmental factors is readily applicable to benign non-state actors as well.

386 Kristan J. Wheaton, email to William N. Reynolds, 15 November 2007. Cited hereafter as Wheaton, email to Reynolds, 15 November 2007.

Table 5. Advantages and Disadvantages of Each Method

Role Potential Spectrum Analysis

Advantages	Disadvantages
Created a standardized foundation to measure environmental influences on the roles of non-state actors to produce comparable findings across Sub-Saharan Africa states.	The major indices in unaltered states are unsuited for measuring the role of non-state actors.
Created a prediction model for the roles of non-state actor in a country.	The matrix required several weeks to finish, and took time away from starting and completing other types of analysis.
The state centric environmental approach allowed the analysts to effectively consider key factors across the entire socio-political environment of individual Sub-Saharan African countries.	Would be difficult for other analysts without statistical background to successfully complete.
Able to represent both government sanctioned and extra government role potentials for non-state actors.	
Guided further efforts to efficiently identify and assess significant indicative characteristics of the socio-political environments in Sub-Saharan Africa.	

Geospatial Analysis

Advantages	Disadvantages
Created a visual representation allowing the team to identify patterns and correlations among different sets of data, particularly with the statistical matrix.	Much of the available information on NGOs, terrorist groups, and businesses was not in English, which made it difficult to collect a larger sampling of the information.
Independent from other analyses.	The team limited the NGO map to development, women's rights, HIV/AIDS, human rights, and environmental issues. This was not an exhaustive list of all the different NGOs operating within the region.
Information retrieved from uniform databases; not truly random, but relatively objective sample.	
Simple and easy to do with the information being imported from Excel spreadsheets directly into community walk.	

Table 5. Advantages and Disadvantages of Each Method (Continued)	
Analysis of Competing Hypotheses	
Advantages	*Disadvantages*
Able to detect an increase or decrease in the roles of non-state actors.	Dependent upon completion of the matrix analysis.
Uniform sources throughout.	Evidence selected tends to be subjective.
Easy for team to complete (received in-class instruction on this type of analysis).	Susceptible to confirmation bias and the anchoring effect.

Source: Mercyhurst College Institute for Intelligence Studies, "Teams' Non-State Actors Process and Methodology Report," Wikispaces.com, URL: <https://nonstateactorsafrica.wikispaces.com/Process+and+Methodology>, accessed 2 March 2008. Cited hereafter as MCIIS, "Process and Methodology."

the sample size, then, we can have more rather than less confidence about the broader conclusions and are likely right about the overall picture.[387]

Such a conclusion is significant for the larger IC. Through this still-experimental application of methods and models we may be able to develop a means of increasing the accuracy of intelligence sensemaking when time is short and the scope is large.

Application: A Multi-Disciplinary Workshop on Non-State Actors

In February 2008, LSS brought together a methodologist, an economist, a political scientist, two psychologists, an anthropologist, and a computer scientist to test the *Warlords Rising,* open-systems approach against a real-life problem: the roles of three sets of non-state actors (representing three major political groups) in Iraq's Anbar Province: Shi'a militia, Sunni sheiks, and Al Qaeda in Iraq (AQI). Participants used four frameworks to consider the environment within which the three groups exist and operate.

- *Points of segmentation* are the boundaries or borders between and among groups of people, where the degree of disagreement on

---

[387] Wheaton, email to Reynolds, 15 November 2007.

issues is indicated numerically.[388] Points of segmentation can track inherent characteristics such as gender or ascribed cultural differentiators such as Sunni or Shiite. The set of points distinguishes one individual or group from another and identifies possible points of cooperation and conflict that can be exploited. Specific values for points of segmentation are derived from an expert assessment of the strength of the actors' expressed attitudes, reinforced by observable behavior. They distinguish one individual or group from another, and identify the points most suitable for exploitation by the protagonist. Numbers were elicited from experts, calibrated against one another for consistency, and used to quantify expert consensus. Computer modeling by Least Squares addressed five pertinent actors: two state-associated, the Iraqi government and the U.S.-led coalition forces; and three non-state-associated actors, the Shi'a militia, the Sunni sheiks, and AQI. The modeling was conducted along five points of segmentation: Arab versus non-Arab; Islamicist versus non-Islamicist; rural versus urban; Sunni versus Shi'a; and strength of tribal allegiance. These issues form a 5-dimensional space — the cells in figure 10 depict the distance between players in this issue space.[389] One especially revealing point of segmentation centers around the question of expected attitudes and behavior in Anbar Province found AQI and Coalition forces on one side, with Shi'a militia, Sunni sheiks, and the Iraqi government on the other side.

As noted by one of the workshop participants, there are three significant implications to segmentation for policy and decisionmakers: First, projects proceed most smoothly *within* a segment. Second, segments that are neglected or discriminated against will push back. Finally, brokers mediating inter-segmental projects need to be viewed as impartial.

---

[388] Points of segmentation is an anthropological term referring to the natural pattern of social divisions within kinship-based societies, in which kinship units form as internally cooperative units. Conflict occurs between segments composed of distinct kinship lines. Segmentation is also used to describe the anthropological analysis of societies into kinship-based units. See Rudolph J. Rummel, *Understanding Conflict and War: War, Power, Peace* (Beverly Hills, CA: Sage Publications, 1991); Benoit Rihoux, "Constructing Political Science Methodology: From Segmented Polarization to Enlightened Pluralism," Joint Chair, Standing Group on Political Methodology. IPSA Conference, Montreal, Canada, 2008; George De Vos and Lola Romanucci-Ross, eds., *Ethnic Identity: Cultural Continuities and Change* (Palo Alto, CA: Mayfield, 1982); and William S. McCallister, "COIN and Irregular Warfare in a Tribal Society," *Small Wars Journal, Blog and Pamphlet, 4 February 2008,* URL: <http://www.smallwarsjournal.com/documents/coinandiwinatribalsociety.pdf>, accessed 31 May 2010.

[389] The technique can be further refined by weighting the different issue axes using expert knowledge. We have omitted this part for brevity.

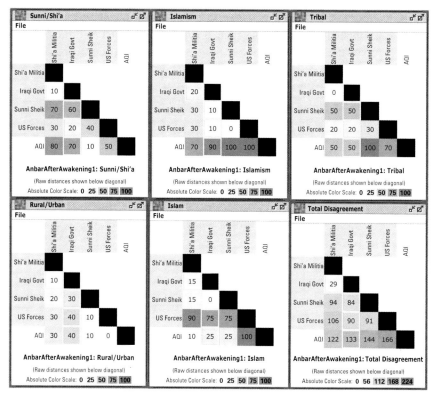

Figure 10. Iraqi Points of Segmentation.

Source: William Reynolds *et alia*, "Social Science Modeling Workshop: Understanding Iraqi Non-State Actors," Workshop Proceedings, Least Squares Software, Albuquerque, NM, 15 February 2008.

- *Prospect theory*, originally developed by Kahneman and Tversky, posits that "people tend to be risk-preferring when facing long shot risks involving significant gains, such as betting on race horses, and are risk averse when facing significant losses: [in other words, when] buying a home or car insurance respectively."[390] Workshop participants concluded that in Iraq, expectations contribute to environments where individuals (and communities) are likely to support or to become non-state actors. Assessment of findings from

---

[390] "Why do People 'Play the Longshot' and Buy Insurance? It's in Our Genes," *Genomics and Genetics Weekly*, 29 January 2010, 184. For a technical description of prospect theory see Daniel Kahneman and Amos Tversky, "Prospect Theory: An Analysis of Decision Under Risk," *Econometrica*, vol. 47, no. 2 (March 1979): 263-291; and Amos Tversky and Daniel Kahneman, "Advances In Prospect Theory: Cumulative Representation of Uncertainty," *Journal of Risk and Uncertainty*, vol. 5 (1992): 297-323.

the application of prospect theory to the Iraqi environment are ongoing, and will be published in forthcoming studies.

- *Institutional interactions* is the name associated with a systematic model that allowed workshop participants to explore the complex roles non-state actors play as they influence (and are influenced by) overlapping institutional capabilities and needs.[391] The participants concluded that even a simple model of institutional networks has enormous complexity—or high entropy—making it a good candidate for a subsequent in-depth modeling project. Due to imposed time constraints, development and application of the modeling was not completed.

- *Morphological analysis* was identified as an additional approach through the institutional interactions method. Morphological analysis considers an entire space of possible implications opening the way for follow-on disambiguation (perhaps using additional multimethodological approaches) in order to abductively and soundly derive the kind of judgments that become useful knowledge.[392]

The workshop participants were unable to formally triangulate the results from the different approaches, also because of insufficient time. Their discussions and modeling, however, supported the *Warlords Rising* thesis: that environment is a critical factor in understanding the emergence and roles of non-state actors. Additionally, the modeling appeared to provide a promise of metrics that, with further development, can be applied against the non-state actor problem.

Critical Assessment: Lessons Learned from the Study of Non-State Actors

No matter what the methodological approach, project participants emphasized that close attention to environmental factors remains a key to understanding non-state actors. Nonetheless, even those approaches that emphasized environmental factors fell prey to certain inadequacies.

---

[391] The application of this model is exemplified in Robert Gibbons and Andrew Rutten, "Institutional Interactions: An Equilibrium Approach to the State and Civil Society," *IQ* online journal, The Institute for Quantitative Social Science at Harvard University. URL: <http://www.iq.harvard.edu/files/iqss/old/PPE/gibbons+rutten.pdf>, accessed 14 May 2010.

[392] Morphological analysis is an approach for considering all the factors and their interrelationships in non-quantifiable, multidimensional problems. Developed by Swiss astrophysicist Fritz Zwicky, morphological analysis begins with the premise that "within the final and true world image everything is related to everything, and nothing can be discarded *a priori* as being unimportant." [Fritz Zwicky, *Discovery, Invention, Research—Through the Morphological Approach* (Toronto, CA: The Macmillan Company, 1969), 44.] It is particularly well suited (and often used) to make sense of non-quantifiable wicked problems.

Both the Mercyhurst and the LSS teams adopted multimethodologically rich approaches to making sense of non-state actors; both quantitative and qualitative methods were employed. While novel insights were generated as has been noted, no means emerged to quantify a specific increase or decrease in non-state actor roles worldwide. Although the multiple methods framed the issues in both complementary and contradictory ways, none, singly or in combination, answered the question of "how much non-state actor 'power' and 'influence' have increased worldwide."[393]

That the members of each study group were working independently supports the contention that the groups were not somehow engaged in a confirmation exercise of each other's work; the conclusions were independently derived. What cannot be determined, however, is the extent—if any—to which specific methods employed in the specific efforts may have impacted conclusions derived from subsequently employed methods (especially in the cases of the MCIIS and LSS teams).

Serendipitous conclusions such as the possible indicator of state instability as illustrated in the case of Kenya by the Mercyhurst team, and the interesting points of segmentation found by the LSS team, suggest further avenues of intelligence research and modeling. The division of Africa reflecting three zones of non-state actor influence was also interesting. But these conclusions, while novel, did not satisfy the initial requirement, a quantification of phenomena associated with change in the power positions of various non-state actors. While all three teams clearly engaged in critical thinking about the issue, as viewed in hindsight, different purposes (specifically as expressed by the NIC/LSS team) led to conclusions different from those of the original, albeit implied, question. That these purposes were different could have been seen as an indicator that the teams were working on different problems within the issue of non-state actors. Coming to a consensus of what the actual tasking was could be expected to have narrowed the divergent results and provided a more specific answer about how the roles of non-state actors have changed.

Changes in the Roles of Non-State Actors: An Alternative View

A systematic review of what was done and not done in the three non-state actor studies provides insights into how critical thinking can combine with multimethodological, mindful sensemaking, to provide a paradigm for 21st Century intelligence creation and its active communication to

---

393 NIC, "Nonstate Actors," 6.

policymakers in a fashion that transcends the Sherman Kent tradition. This review is facilitated by employing ten elements of reasoning.

- **Question:** the beginning question of the NIC-Eurasia Group seminars was, "If non-state actors are emerging as a dominant global force, where is the evidence?" In other words, while there appears to be a consensus that they are a dominant global force, where is the formal evidence? For example, given the premise that we are experiencing an emerging phenomenon, is there evidence that non-State actors wield more power in 2012 (or in 2007 when the original study was made) than they did a decade earlier? A key question, and really the central question, to be answered is first, how does one measure the relative power of non-state actors? A follow-on question to this becomes, "Is such power therefore sufficient to render them dominant global forces?"

- **Purpose:** Determine whether or not there is evidence that non-state actors are emerging as a dominant global force. This problem is one of basic research to determine if the evidence in fact exists. However, the underlying issue of how we might measure relative power must first be conceptualized and addressed.

- **Points of View:** As we consider the original and complementary studies, there are two predominant points of view at issue: first, that of the NIC and its customers—who may believe that non-state actors are an emerging global force and want to quantify this shift in influence and power. The other, unavoidable point of view is that of non-state actors—some of whom would believe they are an emerging force and some who would not believe they are. For example, Al Qaeda, as a non-state actor, might want to believe (and might be justified in believing) that they are a significant global force. On the other hand, a super-empowered individual might believe she or he is not a significant global force (and yet, might be one).

  Each group that considered the issue also reflected different points of view. The NIC-Eurasia Group study focused on three broad categories of "benign" non-state actors: non-governmental organizations, multinational corporations, and super-empowered individuals. Mercyhurst focused on all categories of non-state actors and the LSS workshop focused primarily on violent non-state actors in Iraq. Among other things, this differentiation of what constitutes a non-state actor also reflects different interpretations of the key question. Each group brought its own perspectives about its larger focus as well. For example, both the Mercyhurst and LSS approaches to the issue reflected a strong reliance on structured methods as the means of making sense of issues. The Eurasia group imparted a more intuitive approach by subject matter experts. A more explicit accounting of the points of view embodied at each stage of an ongoing study would help both intelligence producers and their policymaking counterparts maintain a focus on their respective stakeholders' concerns and opportunities.

- **Assumptions:** The use of the term "non-state actor" as an apparent all-encompassing term in the initial problem question and statement presumes an initial understanding and consensus about what is or is not a non-state actor. This is actually inaccurate as the differing foci of the three groups make clear. However, the differences in this case become evident in hindsight although measures could be taken in foresight to at least check the understanding of different groups engaged in collaborative assessments.

  Greater precision of the term non-state actors is needed. Differentiating between benign and non-benign non-state actors is a first step. Subsequent refinements of "benign non-state actors" into nongovernmental organizations, multinational corporations, and super-empowered individuals is also useful. A similar set of distinctions within the set of violent non-state actors is also necessary. Then, a crosscheck among the teams must be accomplished so that consensus on the meaning and use of these terms is achieved.

  Another assumption involves what is meant by the term "dominant global force." Again, both greater precision and clarity is needed in coping with this assumption. One key question is, "Exactly what does dominant global force mean?" One answer to this could be that everywhere on the planet non-state actors are *the* force affecting politics and life. Such a simplified and simplistic view is likely inaccurate, and a range of political process models—among them those of the "rational actor," of "bureaucratic politics," and of "organizational process," need to be parsed.

- **Implications and Consequences:** The consideration of implications and consequences means to anticipate and explore the events that follow a decision, and to put in play especially the interpretive aspect of sensemaking. In the context of non-state actors, it means to explore what happens if non-state actors are (or are not) emerging as dominant global forces and we are right or wrong about their power. Regardless of whether non-state actors are a dominant global force, if their influence is underestimated then surprises can be expected: Some non-state actor is likely to act in a fashion that is completely unexpected and with unanticipated results. On the other hand, overestimating the influence of non-state actors might create self-fulfilling prophecies. If, though, the influence of non-state actors is accurately measured it may be possible to mitigate that influence (where the non-state actors are acting on interests at odds with those of the United States). Alternately, where non-state actors are acting in consonance with the interests of the United States or are able to exploit opportunities put in place to get them to be helpful, the United States fulfills its goals.

Finally, the sensemakers' interpretation of likely actions or events allows the implications and consequences of those actions to be considered, even if absolute prediction is elusive. Here, in the context of collaborative sensemaking through the communication of intelligence to a policymaker, we understand the admonition of Sherman's Kent's contemporary critic, Willmoore Kendall, that intelligence most critically "concerns the communication to the politically responsible laymen of the knowledge which…determines the 'pictures' they have in their heads of the world to which their decisions relate."[394] This vision suggests communication of intelligence as an "insider" rather than offering "intelligence input" at arms length in the Kent paradigm.[395] Kendall faults Kent's equation of wartime and peacetime intelligence, insisting that peacetime intelligence represents a more strategic calling that requires intelligence to consider the course of events as something one must influence, by making what he distinguishes as "contingent predictions."[396] An issue in the case at hand is whether interactions of the United States with non-state actors represent those of war or peacetime intelligence frameworks. Additionally, given the wicked nature of the issue, Kendall's more tailored, "insider" paradigm may prove better in assisting the policy customer in grasping the issue and its contingent predictions.

---

[394] Kendall, "The Function of Intelligence," 550. One might guess that Kendall had in mind the clear depiction of what we now typically call "scenarios."

[395] Richard K. Betts, in *Enemies of Intelligence* (chapter 2), argues for a modulated "politicization" of intelligence. He identifies Secretary of Defense and former Director of Central Intelligence Robert Gates as an effective leading exponent of the "insider" approach. Ultimately, Betts declares, "Taxpayers hire intelligence analysts not to produce truth for its own sake but to produce useful truth," 78.

[396] Kendall, "The Function of Intelligence," 549.

- **Evidence:** What evidence is needed to determine that non-state actors are, and as importantly, are *not* an emerging dominant global force? As we have seen, each group gathered and sifted considerable information on non-state actors, some of it highly relevant to the central question and some not. To the best knowledge of the authors, each of the three groups chose and evaluated evidence only with inductive logic. They did not take advantage of a means, available in particular to a community with robust intelligence capabilities, to deductively eliminate one of the two possibilities.

A seminal essay by an Israeli intelligence practitioner, Isaac Ben-Israel, explains how a technique, viewed in retrospect, would have led the Israelis to dismiss deceptive indicators of Arab preparations and to expect the coordinated attack in October 1973.[397] His finding rests on the idea that intelligence foraging and marshaling capabilities can be used efficiently to focus on a greatly reduced information stream if the reports that support both options (in this case Arab war preparations or exercise) are simply set aside. Reports that are incompatible with either war preparations, on the one hand, or with activity being only an exercise on the other, are few enough to explore with special intelligence means.[398] A focus on detecting evidence of deception in either of those sets would at least bring efficiency to the sensemaking process. In the case of the emerging roles of non-state actors, examining evidence that neither is happening and why would yield alternative views and might also force a disconfirmatory framework allowing better disambiguation of all the hypotheses. The application of this method to intelligence issues is likely not as difficult or inconvenient as practitioners may guess.[399]

---

[397] Isaac Ben-Israel, "Philosophy and Methodology of Intelligence: The Logic of Estimate Process," *Intelligence and National Security*, vol. 4, no. 4 (October 1989): 660-718. Cited hereafter as Ben-Israel, "Philosophy and Methodology of Intelligence."

[398] Ben-Israel, "Philosophy and Methodology of Intelligence," 709.

[399] The challenge of "one-off" prediction is especially inviting to proponents of applying deductive logic to the collection and analysis of evidence. Practitioners who consciously employ this technique can demonstrate to later investigators or inquisitors the logical as well as intuitive steps they took with respect to evidence, such as often occurs in intelligence. In other domains (such as weather forecasting) predictions are constantly updated as inputs to atmospheric models change and as new inputs are revealed. Further, if only an inductive approach to evidence collection and processing is used, when intelligence predictions change, intelligence customers and overseers are often quick, in their own inductive and intuitive way, to criticize either the original or revised results (depending on which set of results they agree with). Thus, a 21st Century approach to forecasting must adopt a constantly updated, deductive paradigm if it is going to be accurate against constantly shifting adversaries and issues.

- **Inferences and Conclusions:** With three different and independent efforts, the challenge lies in ensuring a useful triangulation of the results of those potentially disparate efforts. The Mercyhurst approach (internally triangulated) found that non-state actors, both legal and extralegal, are least effective in authoritarian states. Using Iraq as a case study, the Least Squares workshop demonstrated that within the context of either failed or failing states, expectations and perceptions of the public, or the political environment, are key drivers in anticipating the likelihood of actions by (violent) non-state actors. Strident or acrimonious expression of dissent that arises when domestic and international political/economic issues reinforce each other within the United States and Europe suggest a possible correlation in post-industrial states. This leads to a general conclusion that when expectations are at odds with situational reality, non-state actor activity increases.

  Together, these findings may assist in anticipating the likelihood of future actions by non-state actors. As Schrodt and Gerner note, "political predictions tend to be short-term rather than long-term."[400] Yet, as they also note, effective warning (of a complex humanitarian crisis, for example) requires a fairly long lead-time:

  > Warnings of less than three months provide insufficient lead time for most non-military organizations to react; in other words, the responses to a warning of less than three months will look pretty much the same as a response to a situation that develops without warning.[401]

- **Concepts:** Not only the assumptions, but other concepts as well were in play at multiple levels in the non-state actor case studies. The very notions of "non-state actor" and the ideas of democracy, authoritarianism, and anarchy needed clarification, ideally through well-grounded, empirical as well as theoretical research, to ensure common understanding.

- **Alternatives:** If non-state actors are not emerging as a dominant global force, then what can we say about their global role? Is their influence staying the same? Is it diminishing? Given a credible means of measuring change in the influence and power of non-state actors, the next step in this study of non-state actors would be to examine hypotheses generated from these alternative questions.

---

400 Schrodt and Gerner, "CHC," 4.
401 Schrodt and Gerner, "CHC," 4.

Such follow-on studies also examine and attempt to make sense of instances where non-state actor influence has waned. As was noted by the Eurasia group, in some situations, pro-Western non-state actors actually shift influence away from themselves and the United States and its allies.

- **Context:** As has been repeatedly noted, non-state actors present both a challenge to U.S. interests and an opportunity for advancing those interests. The U.S. would like to mitigate the challenges and take advantage of the opportunities. How to make that happen in domains and regions of little existing U.S. influence or of waning U.S. and Western influence becomes a key concern as the United States strives to carry out a meaningful global role. Future attempts to make sense of the role of non-state actors may benefit from tapping into the larger context of recent policy-relevant literature on the problem of fragile states in applied academic journals.[402]

Moving Beyond a Proto-Revolution

Microcognition and Macrocognition in the Study of Non-State Actors

There emerge two very general domains of which intelligence professionals must make sense: That of the relatively static, state-based system and that of the much more dynamic non-state actor. Of course, these do not exist in isolation from one another. There are boundaries, interstices, and points of segmentation; there is considerable overlap when one usurps or adopts the actions of the other. Further, the separate domains of domestic and foreign areas of interest and action, embraced by the Kent model of intelligence creation and communication, have been superseded by an indivisible, worldwide web of personal and organizational relationships. Broadly speaking, the "classic" model of intelligence sensemaking largely sufficed and perhaps continues to suffice when issues remain clearly tied to the political entities associated with the Westphalian system of state-based power. However when dealing with non-state actors, a new, revolutionary paradigm becomes essential for making sense of issues as well as their interactions with the states of the other paradigm. In the former, a traditional, intuitive and expert-supported approach was largely adequate. In the case of the latter, as is glimpsed in this case study, a more rigorous approach is required.

---

[402] For example, see Kenneth Menkhaus, "State Fragility as a Wicked Problem," *Prism*, vol. 1, no. 2 (March 2010): 85-100. URL: <http//:www.ndu.edu/press/lib/images/prism1-2/6_Prism_85-100_Menkhaus.pdf>, accessed 14 May 2010.

Such an approach is "macrocognitive" in the terminology developed by those who study naturalistic decisionmaking.[403] In national intelligence terms, practitioners and their customers work in a macrocognitive environment as they manage the uncertainty they face in dealing with wicked problems.[404] Macrocognition, then, includes a focus on process as well as results—what we have labeled mindful, self-reflective sensemaking.

Drawing on Klein *et alia*, we observe that intelligence professionals and decisionmakers traditionally are "microcognitively" focused. That is, like those who follow in the Sherman Kent tradition, they are concerned with solving puzzles, searching, and "estimating probabilities or uncertainty values" for different phenomena of interest.[405] As has been discussed, such an approach still may be suitable for solving tame problems or those of the Type 1 domain. Thus, microcognition describes the reductionist foci of the current intelligence "analysis" paradigm. However, this is not sensemaking, which requires another approach.

The transition or shift to macrocognition requires a focus on "planning and problem detection, using leverage points to construct options and attention management."[406] Elements of the foregoing case study exemplify this strategy. Both the Mercyhurst Role Spectrum Analysis and the Least Squares Points of Segmentation identified potential leverage points that revealed truths about non-state actors, leading to more robust problem detection. A next step would have been to take the triangulated results from all the deployed sensemaking methods and use the results to construct options for dealing with nonstate actors in multiple environments. Such a macrocognitive approach would allow more persistent attention to the *anticipation* of the broad course of events (in this case involving non-state actors), in contrast to a microcognitive focus on *predicting* more isolated and specific future incidents.

Next Steps in Revolutionary Sensemaking about Non-State Actors

The foregoing elaboration of non-coordinated sensemaking activities, even with its limitations, moved beyond the traditional model of intelligence creation. It specifically identified the multiple approaches taken by

---

403 Gary Klein and others, "Macrocognition," *IEEE Intelligent Systems*, vol. 18, no. 3 (May/June 2003), 81. Cited hereafter as Klein and others, "Macrocognition."

404 Pietro C. Cacciabue and Erik Hollnagel, "Simulation of Cognition: Applications," in Jean-Michel Hoc, Pietro C. Cacciabue, and Erik Hollnagel, eds., *Expertise and Technology: Cognition and Human-Computer Cooperation* (Hillsdale, NJ: Lawrence Erlbaum Associates, 1995), 55-73.

405 Klein and others, "Macrocognition," 82.

406 Klein and others, "Macrocognition," 82.

independent teams who used alternative schema and methods that, perhaps unsurprisingly, resulted in a broader understanding of the problem. Much of the multimethod work was based on differing perceptions of the task at hand. Triangulation was largely informal both within and between the groups. Thus, the work met the criteria for a *transitional* intelligence sensemaking project. The participants in all three efforts engaged in critical thinking to one degree or another. All were also mindful of the wicked issue of non-state actors and its significance.

To move farther toward a revolutionary paradigm for intelligence, these approaches need to be formalized beyond the transitional phase provided here into a new paradigm of sensemaking. This does not mean that the structured methods of "science" are to be imposed blindly on an issue.[407] Rather than leading to a scientific approach, this could lead to scientism, wherein meaning accrues only to measurable phenomena for which our understanding rests on hypothesis testing and refutation.[408] Instead, the adoption of the new paradigm for sensemaking depends on bringing into play a cooperative spirit of science and scientific inquiry to the process of intelligence creation and communication. Mindful, critical thinking-based, multimethodological approaches to analysis, synthesis and interpretation are one means of doing this. Additionally, a means needs to be found to ensure that this approach to sensemaking remains rigorous. This becomes the subject of the next chapter.

---

[407] Here, one is reminded of a quote attributed to Secretary of Defense Robert McNamara: "We need to stop making what is measurable important and start making what is important measurable."

[408] Friedrich August Hayek, a Nobel Prize-winning economist, developed the distinction between science and scientism. See Friedrich August Hayek, *The Counter-Revolution of Science: Studies on the Abuse of Reason*, 2nd Edition (Indianapolis, Indiana: Liberty Fund, Inc., 1979), 19-25.

CHAPTER 8
Establishing Metrics of Rigor

Defining Intelligence Rigor

I know the distinction between inductive and deductive reasoning. An intelligence officer is inherently inductive. We begin with the particular and we draw generalized conclusions. Policymakers are generally deductive. They start with the vision or general principle and then apply it to specific situations. That creates a fascinating dynamic, when the intelligence guy, who I call the fact guy, has to have a conversation with the policymaker, who I tend to call the vision guy. You get into the same room, but you clearly come into the room from different doors. The task of the intelligence officer is to be true to his base, which is true to the facts, and yet at the same time be relevant to the policymaker and his vision. That's a fairly narrow sweet spot, but the task of the intelligence officer is to operate in that spot.

> — GEN Michael V. Hayden (Ret.),
> former Director of the Central Intelligence
> Agency and the National Security Agency[409]

Michael Hayden's view that the intelligence officer needs to operate in the "sweet spot" linking intelligence and policymaker cognitive worlds coincides with the aim of the sensemaking paradigm. To bring these two worlds together, intelligence professionals can take advantage of the opportunity to meld their fact-based inductive tendencies with the visionary, deductive model of policymakers through the application of collective rigor to well-conceived questions. This approach allows intelligence professionals to embrace a triangulation on wicked problems from their professional perspective, and to improve their chance to communicate with policymakers whose circumscribed comfort zone may accept or even welcome wicked problems as opportunities to apply their vision to bring about politically rewarding solutions.

---

[409] From a 9 August 2010 interview on national security and U.S. strategy, aired by C-SPAN radio, URL: <http://www.cspan.org/Watch/Media/2010/08/09/Terr/A/36779/Gen+Michael+Hayden+Ret+The+Chertoff+Group+Principal.aspx>, accessed 11 August 2010.

At present most tradecraft for sensemaking triangulation remains intuitive, operating in the realm of tacit knowledge. Thus, part of a revolution in intelligence requires that more formal and explicit means of triangulation be developed. It may be that some existing analytic tradecraft, when conscientiously applied, will improve synthesis and interpretation. Another option is to explore and experiment with new tools for conceptualizing rigor in information analysis, synthesis and interpretation.

Rigor in sensemaking can refer to inflexible adherence to a process or, alternatively, to flexibility and adaptation "to highly dynamic environments."[410] As proponents of the latter approach, Daniel Zelik, Emily Patterson, and David Woods recently reframed the idea of rigor into a more manageable concept of "sufficiency."[411] In the applied world of sensemakers, then, an apt question is: "Were sufficient considerations made or precautions taken in the process of making sense of the issue?" Zelik *et alia* observe that this requires a "deliberate process of collecting data, reflecting upon it, and aggregating those findings into knowledge, understanding, and the potential for action."[412] In order to achieve answers to this question Zelik *et alia* developed an eight-element taxonomy of sufficiency and a trinomial measurement of rigor: Each element was calibrated in terms of high, medium or low rigor. In their examination of information products, an overall score could be computed that, in intelligence terms, would communicate to both the practitioner's management and to consumers the rigor of the crafted intelligence product. Their model of the metric is shown in figure 11 and its attributes are reproduced in table 6.

410 Daniel J. Zelik, Emily S. Patterson, and David D. Woods, "Understanding Rigor in Information Analysis," in Kathleen Mosier and Ute Fischer, eds., *Proceedings of the Eighth International NDM Conference*, Eighth International Naturalistic Decision-Making Conference, Pacific Grove, CA, June 2007, 1. Cited hereafter as Zelik, and others, "Rigor."

411 Zelik and others, "Rigor," 1. Note that "sufficiency" differs from "satisficing," by definition a low-rigor process where only enough sensemaking is conducted to get to an initial answer. Indeed, satisficing may be considered the epitome of low rigor.

412 Daniel J. Zelik, Emily S. Patterson, and David D. Woods, "Measuring Attributes of Rigor in Information Analysis," in Emily S. Patterson, and Janet E. Miller, *Macrocognition Metrics and Scenarios: Design and Evaluation for Real World Teams* (Aldershot, UK: Ashgate, 2010), 65. Cited hereafter as Zelik and others, "Measuring Rigor."

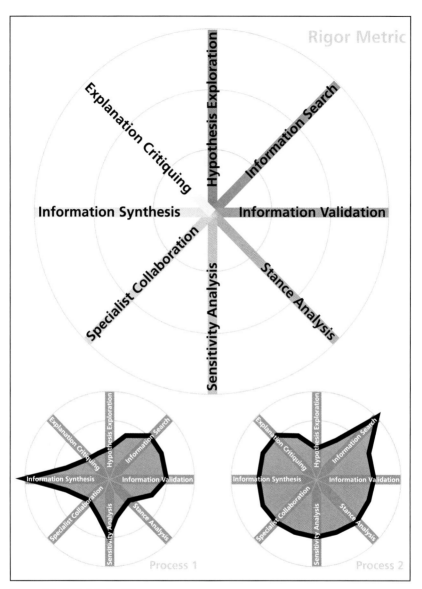

Figure 11. A Metric for Rigor.

Source: Daniel J. Zelik, Emily S. Patterson, David D. Woods, "Modeling Rigor in Information Analysis: A Metric for Rigor," Cognitive Systems Engineering Laboratory, The Ohio State University. Used with permission.

Table 6. Attributes of the Rigor Metric

Hypothesis Exploration describes the extent to which multiple hypotheses were considered in explaining data. In a low-rigor process there is minimal weighing of alternatives. A high-rigor process, in contrast, involves broadening of the hypothesis set beyond an initial framing and incorporating multiple perspectives to identify the best, most probable explanations.

Information Search relates to the depth and breadth of the search process used in collecting data. A low rigor analysis process does not go beyond routine and readily available data sources, whereas a high rigor process attempts to exhaustively explore all data potentially available in the relevant sample space.

Information Validation details the levels at which information sources are corroborated and cross-validated. In a low-rigor process little effort is made to use converging evidence to verify source accuracy, while a high-rigor process includes a systematic approach for verifying information and, when possible, ensures the use of sources closest to the areas of interest.

Stance Analysis is the evaluation of data with the goal of identifying the stance or perspective of the source and placing it into a broader context of understanding. At the low-rigor level an analyst may notice a clear bias in a source, while a high-rigor process involves research into source backgrounds with the intent of gaining a more subtle understanding of how their perspective might influence their stance toward analysis-relevant issues.

Sensitivity Analysis considers the extent to which the analyst considers and understands the assumptions and limitations of their analysis. In a low-rigor process, explanations seem appropriate and valid on a surface level. In a high-rigor process the analyst employs a strategy to consider the strength of explanations if individual supporting sources were to prove invalid.

Specialist Collaboration describes the degree to which an analyst incorporates the perspectives of domain experts into their assessments. In a low-rigor process little effort is made to seek out such expertise, while in a high-rigor process the analyst has talked to, or may be, a leading expert in the key content areas of the analysis.

Information Synthesis refers to how far beyond simply collecting and listing data an analyst went in their process. In the low rigor process an analyst simply compiles the relevant information in a unified form, whereas a high-rigor process has extracted and integrated information with a thorough consideration of diverse interpretations of relevant data.

Table 6. Attributes of the Rigor Metric (Continued)
Explanation Critique is a different form of collaboration that captures how many different perspectives were incorporated in examining the primary hypotheses. In a low-rigor process, there is little use of other analysts to give input on explanation quality. In a high-rigor process peers and experts have examined the chain of reasoning and explicitly identified which inferences are stronger and which are weaker.

Source: Zelik *et alia,* "Rigor," 4.

The tradecraft underlying these attributes assesses the domains of intelligence foraging and intelligence sensemaking (analyzing, synthesizing, and interpreting) as described here. The study by Zelik *et alia* reveals that what might at first glance be considered the application of a low level of rigor really is merely a function of a varying distribution of rigor applied among the attributes. In their study some notable differences appeared in the absolute scores on the rigor scale as different products were scrutinized. However, the more interesting cases involved notable swings in specific attribute scores. Thus, in a test scenario assessing the work of two practitioners (see figure 11), one with a predetermined "high rigor" score and the other with a predetermined "low rigor," the fact that the high-rigor practitioner (Process 2) had a high score for *information search* and a medium score for *information synthesis* whereas the low-rigor practitioner (Process 1) had a medium score for *information search* and a medium score for *information synthesis* is noteworthy.

This distinction suggests a profound insight, namely that *information search* is perceived as more highly valued than *information synthesis*. In the case above, a definitive judgment about this insight cannot be made, as the other attributes of the two assessments were not identical. Still, it seems clear that at least the participants in the study were still wrestling with a consideration formally discussed by Richards Heuer, Jr. (and many others before and since—including Moore and Hoffman above): How much information is necessary for effective sensemaking?[413] The observations may also reflect the "data-centric" culture of the participants. Finally, it offers a means to counter a tendency or preference among "Google Generation" foragers, as described in Chapter 4, for broad but shallow searches. A next step in examining the metric is to apply it to an assemblage of intelligence assessments. An illustration of how this might work using the case study of the previous chapter appears below.

---

413 Heuer, *Psychology,* Chapter 5. See also, Slovic, "Behavioral Problems."

Assessing Sensemaking Rigor in Studies of Non-State Actors

Criteria	NIC-Eurasia Group	Mercyhurst	LSS
Hypothesis Exploration describes the extent to which multiple hypotheses were created and considered in explaining data.	*Low*	*High*	*High*
Information Search relates to the depth and breadth of the processes used in foraging and harvesting data.	*Medium*	*High*	*High*
Information Validation expresses the level to which information sources are corroborated and cross-validated.	*Medium*	*Medium*	*Medium*
Stance Analysis is the evaluation of data with the goal of identifying the stance or perspective of the sources and their placement into a broader context of understanding.	*Medium*	*Medium*	*High*
Sensitivity Analysis considers the extent to which the analyst considers and understands the assumptions and limitations of their analysis.	*Medium*	*High*	*High*
Specialist Collaboration describes the degree to which an analyst incorporates the perspectives of domain experts into their assessments.	*High*	*Low*	*High*

Table 7. Comparative Scores of the Three Efforts Examining the Roles of Non-State Actors

Criteria	NIC-Eurasia Group	Mercyhurst	LSS
Table 7. Comparative Scores of the Three Efforts Examining the Roles of Non-State Actors (Continued)			
Information Synthesis refers to how far beyond simply collecting and assembling data an analyst progressed.	*High*	*High*	*High*
Explanation Critique is a measure of collaboration that captures how many different perspectives were incorporated in examining the primary hypotheses.	*Medium*	*Medium*	*High*
Team Scores (Each low = 1; medium = 2; high = 3)	17	19	23

Sources: Zelik *et alia*, "Rigor," 4; Author.

The case study in the previous chapter affords an opportunity to demonstrate how this rigor metric functions when applied against the work of real intelligence professionals—or in this case, teams of intelligence professionals. In this example the informed judgment of the author leads to three individual rigor metric diagrams as well as a composite; finally, a table summarizes the results.[414] Applying a scoring metric of 1 point for a Low Score, 2 points for a Medium Score, and 3 points for a High Score yields a range of 8 to 24 possible points. For the reviewers, when no evidence pointed clearly to a low or high score, a default position of medium was presumed.

---

[414] The author is indebted to Russell Swenson who reviewed the assessments (and agreed with the scoring of the author). Numerical scores were assigned at his suggestion. It should be observed that this assessment is not intended to be authoritative. Rather it is provided as an example of how rigor can be assessed.

Rigor and the NIC-Eurasia Group Effort

The NIC-Eurasia Group effort (summarized in figure 12 and column one of table 7) garnered the fewest points of the three groups. **Hypothesis Exploration — Low:** The NIC-Eurasia Group memorandum noted that non-state actors are of interest "because they have international clout, but are often overlooked in geopolitical analysis." The implicit but demanding questions of why and how much non-state actor "power" and "influence" have increased worldwide were not answered, nor was a time frame established. This failure to broaden the hypotheses beyond the initial framing of the issue led to a lack of incorporation of multiple perspectives to identify at least "best," and perhaps most probable answers to these questions. **Information Search — Medium:** Based on the memorandum there was no evidence of a "high rigor" or exhaustive information search. **Information**

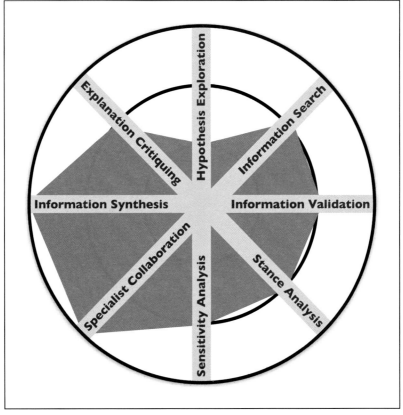

Figure 12. The Rigor of the NIC-Eurasia Group Study of Non-State Actors.
Source: Author, based on Zelik *et alia*, "Rigor."

140

Validation — Medium: Convergence of evidence was apparent in the NIC-Eurasia Group desktop memorandum. **Stance Analysis — Medium:** This was found to be a process-related metric and the desktop memorandum did not adequately reveal the process undertaken. However, some evidence exists that consideration of the source backgrounds took place. **Sensitivity Analysis — Medium:** While the NIC-Eurasia Group's explanations went beyond the "surface level," no obvious evidence was presented of an explicit strategy to consider the strength and sensitivity of explanations about non-state actors. **Specialist Collaboration — High:** Such a score was expected. Eurasia Group is an organization of experts; this is one of its values to the community (and others). Leading experts provided their informed opinions about the roles of non-state actors. **Information Synthesis — High:** The NIC-Eurasia Group desktop memorandum presented integrated information with a thorough consideration of the underlying evidence such as the conclusion that "state-owned enterprises" posing as NGOs often are a front for advancing the interests of a modernizing state-government. **Explanation Critique — Medium:** A discussion process presumes the examination and critique of each participant's chains of reasoning. Since discussions were a part of the sensemaking process undertaken by the NIC-Eurasia Group participants, this form of collaboration was present. However, the degree to which this occurred could not be determined from the desktop memorandum, resulting in a score of medium.

Rigor and the Mercyhurst Effort

The work of the Mercyhurst College students earned a score midway between the other two groups. Their **Hypothesis Exploration** was judged to be **High**. Their considerations of the issue clearly went beyond their initial framing of the issue. Similarly, their **Information Search** also scored **High**. A broad range of evidence was employed in making sense of non-state actors in Africa.[415] On the other hand, while the information convergence necessary for the "non-state actor potential spectra" was indicative of high rigor, an informal non-systematic approach resulted in **Information Validation** being assigned a **Medium** score. **Stance Analysis** also was scored as **Medium**. As was the case with the NIC-Eurasia Group effort, this process-related metric could not be further evaluated. With regard to **Sensitivity Analysis**, the Mercyhurst effort was scored as **High:** The participants focused on the non-state actors themselves as well as the means by which they could be assessed. Such a focus leads naturally to a consideration of the

---

[415] This was not unexpected. A finely honed skill of participants in the Mercyhurst College Institute for Intelligence Studies program is that of effective, broad and deep search.

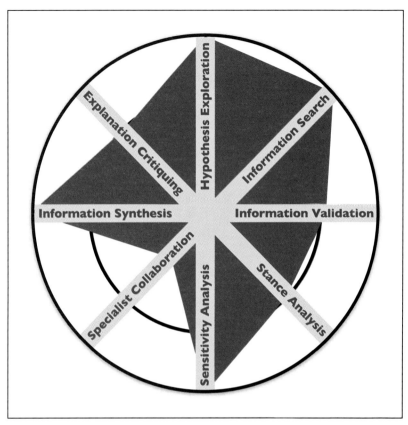

Figure 13. The Rigor of the Mercyhurst College Students' Study of Non-State Actors.
Source: Author, based on Zelik *et alia*, "Rigor."

strengths and weaknesses of explanations. The Mercyhurst students relied on published data. No evidence of consultation with external experts was evident. While this was to be expected given the demographics of a student team, it nevertheless led to a score of **Low** for the **Specialist Collaboration** metric. By contrast, **Information Synthesis** received a **High** score. The consideration of diverse interpretations of the data led the Mercyhurst students to anticipate that Kenya had stability issues well in advance of the breakout of politically related violence. This was unexpected and possibly unanticipated elsewhere. Finally, the **Explanation Critique** metric was scored a **Medium**. The review by the mentoring faculty member and fellow students provided a level of critique that, while valuable, did not meet the standard set by a full peer and expert review of inferences and conclusions, which is necessary for a high score in this category.

Rigor and the LSS Effort

The LSS social science study of non-state actors scored the highest of all three groups, earning a high in each metric save one, **Information Validation**, where they scored a **Medium**. In this case, while converging information was employed to cross-validate source accuracy for the evidence closest to the areas of interest, a systematic approach for doing this was not evident, resulting in the lower score.[416] Overall the LSS effort went well beyond the initial framing of the issue, resulting in a **High** score for **Hypothesis Exploration**. Similarly, their **Information Search** considered every source they could access within their available resources. It was clearly evident that

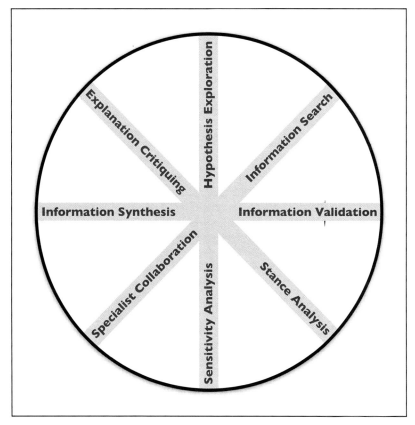

Figure 14. The Rigor of the Least Squares Social Science Study of Non-State Actors.
Source: Author, based on Zelik *et alia*, "Rigor."

---

416 This determination may be overly stringent, as the LSS Stance Analysis (a related metric) was rated as high.

this group conducted rigorous research into source background, facilitating an understanding of how their respective points of view might influence their stances on the issues of violent non-state actors. The use of a formal model of critical thinking added considerable rigor to both **Hypothesis Exploration** and **Sensitivity Analysis**. Furthermore, their conversations specifically focused on the strength of explanations. The use of multiple points of segmentation also captured the diverse alternatives of this complex (or "wicked") issue. Since the entire team was made up of specialists, their score in **Specialist Collaboration** should come as no surprise. Similarly, the involvement of diverse social scientists as well as the involvement of external peers foreordained that the **Explanation Critique** would be rigorous. All the individuals brought differing perspectives as they identified the strengths and weaknesses of each other's inferences and conclusions. Finally, an explicit multi-methodological approach forced consideration of diverse interpretations of the evidence—a highly rigorous example of **Information Synthesis**.

Observations and Discussion

It is no accident that the traditional means by which assessments of such issues are created, as evidenced by the NIC-Eurasia Group effort, resulted in a relatively weak score, whereas the highly rigorous, critical-thinking based, multimethodological effort by a collaborative team of diverse experts led to a relatively high score (a comparison highlighted by figure 15). It should also be noted that absent the LSS effort, a composite effort by the NIC-Eurasia Group and Mercyhurst teams would have come close to the score attained by the LSS group (21 versus 23 points). Here again, the utility of this metric becomes evident. Teams that complement each others' strengths and mitigate each others' shortcomings can be assigned the same general problem, with the expectation of a more rigorous composite effort and more suitable result. As has been illustrated here, the results of individual efforts and combined efforts can be assessed for their rigor as a guide to the purposeful improvement of sensemaking. Specific feedback can be provided so that efforts can be adjusted.

Another advantage of graphic analysis using Zelik *et alia's* metric is that more information can be clearly conveyed. For example, in examining the composite efforts of all three groups of participants in the non-state actor study, it is evident that Information Validation could have been improved through the use of a more rigorous systematic approach, ensuring that the sources were deemed valid and "closest to the areas of interest."[417] In 2005,

---

417 Zelik and others, "Rigor," 4.

144

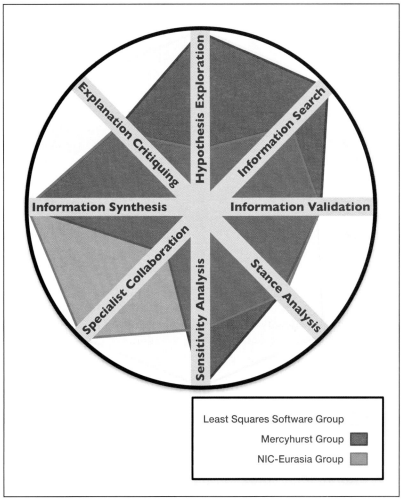

Figure 15. The Composite Rigor of the Three Studies of Non-State Actors.
Source: Author, based on Zelik *et alia*, "Rigor."

an official report on the IC's performance specifically called for improvement in this area.[418]

Despite this call for improvement, there appear to be several reasons why information is often not fully validated in intelligence work. First,

---

[418] The Commission on the Intelligence Capabilities of the United States Regarding Weapons of Mass Destruction (WMD) *Report to the President of the United States* (Washington, DC: Government Printing Office, 31 March 2005), 15, and 371-372. The report recommends that asset validation procedures (a part of information validation) be systematized and standardized.

validation is difficult and the intelligence professional may decide the result is not worth the effort, or that initial conditions *suggest* validity. "Information hubris"—arising when similar information has without negative consequence been presumed or found to be valid, may compound this effort. Wishful thinking and belief in the infallibility of the source are other factors that may contribute to this pathology. Finally, information uncertainty may allow it to resist validation. Unfortunately any or all of these can lead to intelligence errors and failures, suggesting that information validation, despite its inclusion in the rigor metric itself, may require a transcendent application of rigor.

In applying the metric it becomes apparent that some disambiguation is necessary between several of the individual considerations. The differences between high rigor assessments involving Stance Analysis and Sensitivity Analysis at first glance appear to be unusually subtle, suggesting a need for an explanatory critique as part of the standard process assessment. Additionally, because several of the specific metrics are process-related, assessors need to be present to observe the process or otherwise have access to appropriate and sometimes-scarce process-associated materials. Alternately, a formal means for capturing applied (and omitted) sensemaking processes needs to be developed.

In developing the rigor metric, Zelik *et alia* note that it is "grounded largely in the domain of intelligence analysis."[419] Looking at the metric from a generalizing point of view, Zelik *et alia* are interested in whether it can be broadened to other disciplines such "as information search by students working on educational projects, medical diagnosis using automated algorithms for x-ray interpretation, and accident investigation analyses."[420] For those of us within the domain of intelligence, however, that this model "emerged from studies of how experts 'critique' analysis (rather than how experts 'perform' analysis)" is a strength.[421] The rigor metric has been *empirically*, if tentatively, shown (according to Daniel Zelik) to "reveal [some of the] "critical attributes to consider in judging analytical rigor" in intelligence sensemaking.[422] In so doing, Zelik and his colleagues also validated the usefulness of the model within the intelligence domain.

---

419 Zelik and others, "Measuring Rigor," 77.

420 Zelik and others, "Measuring Rigor," 77-78. Some of these domains contain applicability to intelligence. Given a tendency for shallow searches by younger people, then a metric for rigor that offers guidance may facilitate the "deep diving" necessary for adequate information foraging.

421 Daniel J. Zelik, email to the author, 15 October 2010. Cited hereafter as Zelik, email, 15 October 2010.

422 Zelik, email, 15 October 2010.

While the rigor metric requires further development, it offers an effective means of assessing the processes underlying intelligence assessments. The rigor metric provides a means of assessing the processes of intelligence sensemaking, allowing managers of intelligence professionals to ascertain whether or not more work is required before the results of the sensemaking are communicated. Additionally, it offers a practical technique to facilitate communication between practitioner and consumer; that is, to promote *sensemaking*. A danger inherent in any process of making sense of an issue exists when the "process is prematurely concluded and is subsequently of inadequate depth relative to the demands of [the] situation."[423] Finally, the metric suggests other sensemaking considerations that, if used, may help reduce inadequacies in the consideration of intelligence issues. This may lead further to an honest reduction in sensemakers' and their customers' uncertainty about issues being examined. In sum, the metric for rigor developed by Zelik and his colleagues is itself an objectification of what they find rigor to be and warrants further study.[424] Zelik, *et alia* note that the metric "represents a current iteration of an ongoing direction of exploration."[425] Such studies, rigorously making sense of rigor, would be yet another example of the model for intelligence sensemaking that has been championed in this book. As Zelik *et alia* conclude, "the concept of analytical rigor in information analysis warrants continued exploration and diverse application as a macrocognitive measure of analytical sensemaking activity."[426] But what does it mean if such a model is adopted, or not adopted? The final chapter examines the implications of either outcome.

---

[423] Zelik and others, "Measuring Rigor," 65.
[424] Zelik and others, "Rigor," 5.
[425] Zelik and others, "Measuring Rigor," 78.
[426] Zelik and others, "Measuring Rigor," 78.

CHAPTER 9
In Search of Foresight: Implications, Limitations, and Conclusions

Considering Foresight

We turn in conclusion to a discussion of the purpose of mindful, critical sensemaking for intelligence. The discussion may be best framed by pertinent questions: To what end is intelligence intended? In other words, intelligence professionals and their overseers critically ask "knowledge of what?" and, secondly, "knowledge for whom?" One answer to these questions is embodied in the concept of foresight: Intelligence knowledge advises policymakers and decisionmakers about what phenomena are likely precursors of events of interest before they occur. Such foresight—in light of the discussions in this book—does not entail specific predictions. Rather, it allows us to anticipate a range of alternative event sequences.

Foresight informs policy and decisionmakers about what could happen so that those individuals can improve the quality of their decisions. Done mindfully, its vision shifts and evolves apace with the phenomena about which it makes sense. Done wisely in such a manner as presented here, it augments the vision of leaders, enabling mobilization and discouraging two traits that often handicap visionaries: recklessness and intolerance. Done rigorously, it cannot be accused of failing to be imaginative. This prospective approach contrasts with the current practice and paradigm for intelligence production.

Reprising Anthony Olcott's observations reminds us again that Sherman Kent's vision of national intelligence—his "intellectual genetics"—has predominated in U.S. national intelligence since the late 1940s.[427] That it did so was because it "served the United States extremely well for a long time. However, as happens when environments undergo dramatic change, successful adaptations for one environment can prove to be much less efficacious—perhaps even fatal—in a new environment."[428] In the Kent tradition, as has been noted (and is summarized in figure 16), intelligence knowledge of "analyzed" issues becomes and tends to remain disaggregated into constituent parts—oriented, as Treverton notes, toward solving isolated "puzzles" rather

---

[427] Olcott, "Revisiting the Legacy," 21.
[428] Olcott, "Revisiting the Legacy," 21.

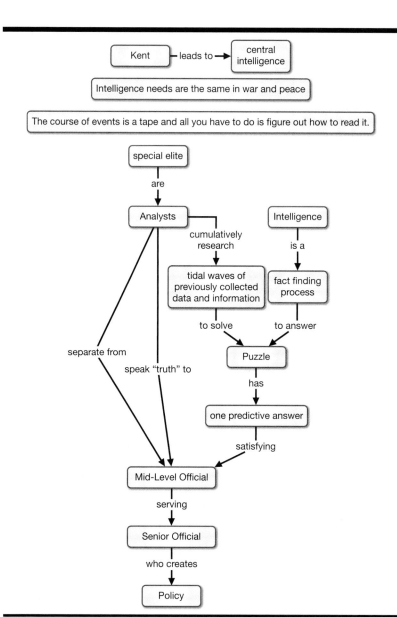

Figure 16. Conceptual Comparison of Kent and Kendall. These diagrams track the potential nodes that signal the differences in the visions of Sherman Kent and Willmoore Kendall. The diagram to the right also traces a revolutionary, Kendallian pathway toward the creation and communication of strategic intelligence.

Sources: Author; based on Jack Davis, "Analysis and Policy: The Kent-Kendall Debate of 1949," *Studies in Intelligence*, vol. 36, no. 2 (1992); Kendall, "Function of Intelligence;" Kent, *Strategic Intelligence*; Olcott, "Revisiting the Legacy;" and Treverton, "Estimating."

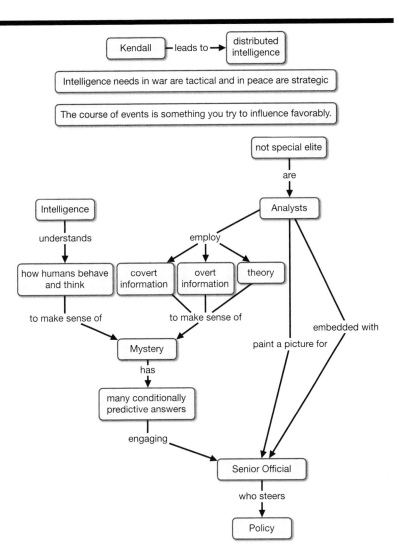

than the more holistic "mysteries" of the intelligence world. On the other hand, we have discussed how Kendall's competing idea, that intelligence knowledge should "paint a picture" (by way of a macrocognitive, holistic approach) for the policy maker as a fellow "insider," is consistent with a model of intelligence where the predominant method and motive of intelligence sensemaking is through aggregation and the articulation of a fact-based "vision" recognizable by national-level policymakers (figure 16). Even in an operational military scenario, where isolated, specific facts are essential to successful employment of mission knowledge, a larger intelligence sense of who the ultimate commanders are and why they are doing what they're doing, remains the essence of useful, foresightful strategic knowledge—also known as national intelligence.

Implications

Creating intelligence as presented here is a mindful process of sensemaking, encompassing the activities of planning, foraging, marshaling, understanding, and communicating. It is critical of itself and the means that are employed in bringing it about. It lies within the largely overlooked Kendallian vision of what national intelligence ought to accomplish. This approach allows a focus on better and worse solutions, and anticipation of likely futures, instead of a more narrow focus on right and wrong answers in an intellectual environment trained on predictive and specific warning. It can make sense of wicked problems.

By contrast, intelligence as it is currently practiced is still somewhat akin to the practice of medicine in the 14th Century. In medieval medicine, herbs and poultices were applied without (from a modern perspective) an understanding of why they might work. If the patient survived, then the method worked. If not, then from the medieval perspective, God willed it otherwise. Intelligence practitioners find themselves in a similar situation. They often do not know why they do what they do, only that the last time, it "worked"—or that it is an "accepted practice." They do not acknowledge that they have "forgotten" all the times it did not work. Yet, intelligence practitioners who would wear the "professional" label need to know what they are doing and why.[429]

One means set forth for "improving intelligence" is to capture the processes by which sense is made of an issue. It is certainly true that imposing audit trails is a critical step because they encourage process improvement

---

[429] It is pertinent to note that suggesting, "God willed it otherwise," is unacceptable for justifying either an intelligence error or an intelligence failure.

in the light of serious errors, and stimulate repetitive analysis, synthesis, and interpretation for validation in the full course of sensemaking. However, auditing trails remain inadequate when the community cannot understand from an epistemological point of view what does and does not work and in what situations.

The major intelligence failures of the first years of the present decade, as well as repeated failures over at least six decades, demonstrate what happens when there is a formal failure to synthesize and interpret beyond what is popularly believed or even to recognize that a situation exists that requires new synthesis and interpretation. A popular hypothesis is that tradecraft can minimize the likelihood of such failures of imagination. Yet this hypothesis remains untested except in some anecdotal cases which, given the Type 2, wicked nature of the intelligence issues now often faced by the community, is inadequate. Indeed, as Grove *et alia* discovered in their psychological meta-study (discussed above), technological approaches (a structuring of sorts) provided accurate solutions less that half the time.[430] Thus, a community effort to research, test, and evaluate tradecraft remains a need if for no other reason than so the IC can understand what approaches to foresightful sensemaking are likely to be useful and in what circumstances. The above-mentioned failures (and others) show that the typical "analytic" paradigm that remains in place leads to failures of imagination in policy circles as well as in intelligence cloisters.[431] The most recent failures point out that the community still does not understand how desperately it needs to make sense.

Limitations

It should be noted that a tradecraft of mindful understanding does not guarantee accurate findings. Any of the components of sensemaking can be done poorly yet "correct" answers can be reached. Disaggregating phenomena can be done well yet yield faulty results. Synthesis and interpretation of analyzed phenomena can still lead to faulty conclusions. However, analysis, synthesis and interpretation within the framework of appropriately applied, multimethod tradecraft does guarantee more rigorous sensemaking. As illustrated in the case study presented above (chapter 7), greater rigor can reveal the hidden discrepancies in an intelligence problem. In the example contained in chapter 5, it can be seen that increased rigor may reveal whether the discrepancies comprise self-generated erroneous conclusions or whether one has been manipulated by a deceptive adversary. A rigorous

---

[430] Grove and others, "Meta-Analysis," 25.

[431] *The 9/11 Report*, 339. The four failures are discussed in depth in subsequent sections of the commission's report.

multimethodological approach will increase the reliability and validity of the findings. Taken together, the entire process certainly will increase the intelligence practitioner's understanding of the issue, allowing more informed communication of associated knowledge to a decisionmaker. This may lead to increased "accuracy." At best, errors will occur with less frequency, or at least they are more likely to be caught.

Conclusions

As has been noted repeatedly in this book, many 21st-Century intelligence issues are wicked problems: They are ill-defined and poorly understood issues with multiple goals that must be made sense of within severe time constraints; the stakes and risks are high and there exists no tolerance for failure. As a means of increasing situational awareness, merely creating mindfulness about such complex issues falls short. On the other hand, a mindful *sensemaking* approach to situational awareness accomplishes more by enabling the intense, holistic scrutiny of a complex developing scenario, as suggested in the case study in chapter 7. This macrocognitive approach ensures that the knowledge created also evolves.

Schwartz and Randall believe that to be successful against strategic surprise—a goal the IC seeks—organizations must be both imaginative and systematic.[432] This is so, because "[one] cannot foresee strategic surprise without being imaginative, but the results will not be believable without being systematic."[433] If intelligence is to rise above the noise and get the attention of policy and then be acted upon it must be both. A critical, mindful process of sensemaking offers a means for this to occur. As we have seen, it covers the issue broadly, takes into account its complexity, is systematic and rigorous. It offers the best means currently understood for making sense of what is known and knowable.

There is and will always be information of which intelligence professionals are unaware. There is certainly information that does not exist because its precursors have not yet occurred. There also is information that exists but is unavailable to the intelligence sensemaker and their associated decisionmakers. Any of this information can change the conclusions reached in making sense of it. In short, intelligence professionals currently make judgments and will continue to make judgments that, while they appear valid at the time, differ from what ultimately occurs, either because of omission or commission. For evidence of this one need not look further than at the differing

---

432 Schwartz and Randall, "Ahead of the Curve," 96-100.
433 Schwartz and Randall, "Ahead of the Curve," 97.

judgments contained in the U.S. National Intelligence Council's series of global trends summaries published every five years since 1995.[434] These documents, which look outward 15 years, perhaps are better seen as expressing the issues and concerns of the times in which they are written, suggesting the validity of Taleb's point that to predict the future one must already be in the future.[435]

The spread and use of the sensemaking paradigm, wherein intelligence professionals move from a reductionist, analytic view to formally include the other elements of sensemaking—synthesis and interpretation—increases the value of intelligence findings. Engaging in macrocognitive sensemaking, with its multiple foci of (re)planning for problem detection, use of leverage points in constructing options, and management of both attention and uncertainty, further creates an approach that will improve communication with consumers of intelligence.[436] As Zukav's epigraph at the beginning of this book reminds us, "[nonsense] is nonsense only when we have not yet found that point of view from which it makes sense."[437] Such an inclusive process provides that point—or rather *points*—of view; nonsense is transformed into vital, strategic, and foresightful knowledge facilitating better decisions by leaders.

Some elements of the IC have begun to establish "tradecraft" cells or similar centers focusing on the practice and process of intelligence. Typically, however, they have not transcended the "analysis" paradigm and remain mired in the uncertainty reduction that is characteristic of the practice of the past sixty-plus years. At best, they are not yet able to transform a professional workforce. Pierre Baumard in 1994 anticipated the persistence of this state of affairs: "[Individuals] act on incomplete and variously reliable information. Caught by approaching dead-lines, surrounded by urgency, individuals seek the simplest means to reduce complexity according to the criteria on which they will be *locally* judged."[438] Although some might gain hope from the renewed interest in tradecraft, the fact that the same errors are being repeated by a new generation of intelligence professionals certainly should lead one to doubt that an underlying and necessary paradigm shift has occurred. That is, the change is not occurring because before the "opponents" die off they are

---

434 The most recent version of the NIC forecasts is *Global Trends 2025: A Transformed World*, URL: <http://www.odni.gov/>, accessed 12 December 2008.

435 Taleb, *The Black Swan*, 173.

436 Klein and others, "Macrocognition," 82-83.

437 Gary Zukav, *The Dancing Wu Li Masters: An Overview of the New Physics* (New York, NY: Harper Collins, 2001), 117.

438 Baumard, "From Noticing to Making Sense, 31-32. Emphasis in original.

acculturating a new, younger generation.[439] If the community is to change it cannot wait for another generation (or longer), especially if that new generation is hired and acculturated into the inadequate paradigm of the preceding or present generations.[440]

Achieving this paradigm shift therefore requires revolutionary action by a revolutionary workforce. Aggressive, mindful sensemaking is one pathway to this new paradigm, and may require a different mix of skills and abilities than is currently present. It certainly requires greater, authentic diversity. Considering the present community, one is reminded of Kent's quip, "When an intelligence staff has been screened through [too fine a mesh], its members will be as alike as tiles on a bathroom floor—and about as capable of meaningful and original thought."[441] In contrast, making sense of the 21st Century's intelligence challenges requires as much rigorous, "meaningful and original thought" as we can muster. Sensemaking, as it has been developed here, offers us a means of creating that desperately needed thought.

For the community to remain unchanged is unacceptable: the implications of being wrong are too dire. Developing a validated practice of mindful sensemaking for intelligence, while not a panacea, is a necessary first step in making sense of the nonsense. The results from IARPA's ongoing exploration of sensemaking, if successful, also can be expected to shed light on how this can practically and functionally be accomplished. Both represent a step toward a true professional practice of intelligence that can meaningfully make sense of the national security challenges of the 21st Century.

---

[439] Planck observed, "A new scientific truth does not triumph by convincing its opponents and making them see the light, but rather because its opponents eventually die, and a new generation grows up familiar with it." Max Planck, *Scientific Autobiography and Other Papers*, F. Glynor, trans. (New York, NY: Philosophical Library, 1949), 33-34. Cited in Charles Weiss, "Communicating Uncertainty in Intelligence and Other Professions," *International Journal of Intelligence and CounterIntelligence*, vol. 21, no. 1 (Spring 2008), 78-79.

[440] While beyond the scope of this book, organizing to bring about a "revolution in intelligence affairs" is another important aspect of focusing intelligence for the century in which it finds itself. William Nolte, writing in *Studies in Intelligence*, explores the environment for reorganization. Yet, Nolte also presages the argument of this volume in noting that "we need to focus less on structure and more on behavior." See William Nolte, "Keeping Pace with the Revolution in Military Affairs," *Studies in Intelligence*, vol. 48, no. 1 (Winter 2004), 1-10.

[441] Kent's comment appears in a footnote on page 74 of his *Strategic Intelligence for American World Policy* (1951 and 1966 printings). It is reproduced in Jack Davis' Occasional Paper for the Kent Center at the CIA. Although Kent was reacting to the stringent security policies of the McCarthy era, his remark also applies to all workforce acquisition practices that promote anything but the widest of diversity. See Jack Davis, "Sherman Kent and the Profession of Intelligence Analysis," *The Sherman Kent Center for Intelligence Analysis*, Occasional Papers, vol. 1, no. 5 (November 2002), URL: <https://www.cia.gov/library/kent-center-occasional-papers/vol1no5.htm>, accessed 10 June 2010.

REFERENCES

Anonymous. *For Want of a Nail Rhyme.* URL: <http://www.rhymes.org.uk/ for_want_of_a_nail.htm>. Accessed 10 September 2007.

Argyris, Chris. *Reasons and Rationalizations: The Limits of Organizational Learning.* Oxford, UK: Oxford University Press, 2006.

Bain, Ben. "A-Space Set to Launch this Month." *Federal Computer Week,* 3 September 2008. URL: <http://fcw.com/articles/2008/09/03/ aspace-set-to-launch-this-month.aspx>. Accessed 15 June 2010.

Baron, Joan Boykoff and Robert J. Sternberg, eds. *Teaching Thinking Skills: Theory and Practice.* New York, NY: Freeman, 1987.

Baumard, Philippe. "From Noticing to Making Sense: Using Intelligence to Develop Strategy." *International Journal of Intelligence and CounterIntelligence* 7, no 1 (Spring, 1994): 29-73.

Berkowitz, Bruce. "U.S. Intelligence Estimates of Soviet Collapse: Reality and Perception." In Francis Fukuyama, ed. *Blindside: How to Anticipate Forcing Events and Wild Cards in Global Politics.* Washington, DC: Brookings Institution Press, 2007: 29-41.

Ben-Israel, Isaac. "Philosophy and Methodology of Intelligence: The Logic of the Estimate Process." *Intelligence and National Security* 4, no. 4 (October 1989): 660-718.

Betts, Richard K. *Enemies of Intelligence: Knowledge and Power in American National Security.* New York: Columbia University Press, 2007.

Bialki, Carl. "Swine Flu Count Plagued by Flawed Data." *Wall Street Journal.* Online edition, 23 January 2010. URL: <http://online.wsj. com/article/SB10001424052748704504970 4575019313343580460. html>. Accessed 2 February 2010.

Blight, James G. and David A. Welch, Eds. *Intelligence and the Cuban Missile Crisis.* London, UK: Frank Cass, 1998.

Bouquet Cyril and Ben Bryant. "The Crisis Is Here To Stay. Do You Have The Key To Coping?" *Forbes.* Online edition, 21 April 2009. Online Edition. URL: <http://www.forbes.com/2009/04/21/ stress-coping-mindfulness-leadership-managing-fixation.html>. Accessed 13 January 2010.

Brei, William S., CAPT, USAF. *Getting Intelligence Right: The Power of Logical Procedure*, Occasional Paper Number Two. Washington, DC: Joint Military Intelligence College, 1996.

Brentano, Franz. *Psychology from an Empirical Standpoint*. Antos C. Rancurello, trans. New York, NY: Humanities Press, 1973.

Brentano, Franz. *Psychologie vom empirischen Standpunkte*. Leipzig, DE: Dunker and Humblot, 1874.

Brewer, John and Albert Hunter. *Foundations of Multimethod Research: Synthesizing Styles*. Thousand Oaks, CA: Sage Publications, 2006.

Bryant Ben and Jeanny Wildi. "Mindfulness." *Perspectives for Managers* 162 (September 2008). URL: <http://www.imd.ch/research/publications/upload/ PFM162_LR_Bryant_Wildi.pdf>. Accessed 14 January 2010.

Burton, Robert A. *On Being Certain: Believing You Are Right Even When You're Not*. New York, NY: St. Martin's Press, 2008.

Cacciabue, Pietro C. and Erik Hollnagel, "Simulation of Cognition: Applications," Expertise and Technology: Cognition and Human-Computer Cooperation. In Jean-Michel Hoc, Pietro C. Cacciabue, and Erik Hollnagel, eds. *Expertise and Technology: Cognition and Human-Computer Cooperation*. Hillsdale, NJ: Lawrence Erlbaum Associates, 1995: 55-73.

Campbell, Kenneth J. "Major General Charles A. Willoughby: A Mixed Performance," unpublished paper, URL: <http://intellit.muskingum.edu/wwii_folder/wwiifepac_folder/wwiifepacwilloughby.html>. Accessed 5 January 2010.

Campbell, Kim, ed. *Addressing the Causes of Terrorism: The Club de Madrid Series on Democracy*, Volume 1. Madrid, SP: Club de Madrid, 2005.

Cantor, Norman F. *In the Wake of the Plague: The Black Death & the World it Made*. New York, NY: The Free Press, 2001.

Central Intelligence Agency. "Iraq's WMD Programs: Culling Hard Facts from Soft Myths." Press Release, 28 November 2003. URL: <https://www.cia.gov/news-information/press-releases-statements/press-release-archive-2003/pr11282003.html>. Accessed 9 December 2009.

REFERENCES (Continued)

Chamberlin, Thomas C. "The Method of Multiple Working Hypotheses." *Science* 15 (old series), no. 366 (7 February 1890): 92-96.

Charnov, Eric L. "Optimal Foraging: The Marginal Value Theorem." *Theoretical Population Biology* 9, no. 2 (April 1976): 129-136.

Cheikes, Brant A., Mark J. Brown, Paul E. Lehner, and Leonard Adelman. *Confirmation Bias in Complex Analyses*. Mitre Technical Report, MTR 04B0000017. Bedford, MA: Mitre, 2004.

Christensen, Clayton. *The Innovator's Dilemma*. Cambridge, MA: Harvard University Press, 1997.

Christensen, Karen. "Thought Leader Interview, Chris Argyris." *Rotman: The Magazine of the Rotman School of Management*. University of Toronto, Winter 2008: 10-13.

Cilliers, Paul. "Knowing Complex Systems." In Kurt A. Richardson, ed., *Managing Organizational Complexity: Philosophy, Theory, and Application*. Greenwich, CT: Information Age Publishing, 2005, 7-19.

Clapper, James. "Remarks and Q & A by Director of National Intelligence Mr. James Clapper." Bipartisan Policy Center (BPC) — The State of Domestic Intelligence Reform. 6 October 2010. URL: <http://www.dni.gov/speeches/20101006_speech_clapper.pdf>. Accessed 29 October 2010.

Clayton, James, D. Y*ears of MacArthur, 1945-1964*. Boston: Houghton Mifflin, 1985.

Cloud, John. "Losing Focus? Studies Say Meditation May Help." *Time*. Online edition, 6 August 2010. URL: <http://www.time.com/time/health/article/0,8599,2008914,00.html>. Accessed, 11 August 2010.

Cohen, Robert S., Paul K. Feyerabend, and Marx W. Wartofsky. *Essays in Memory of Imre Lakatos: Boston Studies in the Philosophy of Science*, Volume XXXIX. Dordrecht, HL: D. Reidel Publishing Company, 1976.

Commission on the Intelligence Capabilities of the United States Regarding Weapons of Mass Destruction. Report to the President of the United States. Washington, DC: Government Printing Office, 31 March 2005.

REFERENCES (Continued)

Cooper, Jeffrey. *Curing Analytic Pathologies: Pathways to Improved Intelligence Analysis*. Washington, DC: Central Intelligence Agency, Center for the Study of Intelligence, 2005.

Crenshaw, Martha. "Political Explanations." In Kim Campbell, ed. *Addressing the Causes of Terrorism: The Club de Madrid Series on Democracy*, Volume 1. Madrid, SP: Club de Madrid, 2005: 13-18.

Davis, Jack. "Sherman Kent and the Profession of Intelligence Analysis." *The Sherman Kent Center for Intelligence Analysis*. Occasional Papers 1, no. 5 (November 2002). URL: <https://www.cia.gov/library/kent-center-occasional-papers/vol1no5.htm>. Accessed 10 June 2010.

Davis, Jack. "Introduction — Improving Intelligence Analysis at CIA: Dick Heuer's Contribution to Intelligence Analysis." In Richards J. Heuer, Jr. *Psychology of Intelligence Analysis*. Washington, DC: Center for the Study of Intelligence, 1999.

Davis, Jack. "Analysis and Policy: The Kent-Kendall Debate of 1949." *Studies in Intelligence* 36, no. 5 (1992): 91-103.

Dervin, Brenda. "Sense-Making Methodology Site." URL: <http://communication.sbs.ohio-state.edu/sense-making>. Accessed 12 September 2007.

Dervin, Brenda, Lois Foreman-Wernet, and Eric Lauterbach. *Sense-Making Methodology Reader: Selected Writings of Brenda Dervin*. Cresskill, NJ: Hampton Press, Inc., 2003.

De Vos George and Lola Romanucci-Ross, eds. *Ethnic Identity: Cultural Continuities and Change*. Palo Alto, CA: Mayfield, 1982.

Economist. "Garbage In, Garbage Out." 3 June 1970: 70.

Economist. "Which MBA? 2009 Full-time MBA Ranking." URL: <http://www.economist.com/business-education/whichmba/>. Accessed 15 January 2010.

Eggen, Dan and Walter Pincus. "Ex-Aide Recounts Terror Warnings: Clarke Says Bush Didn't Consider Al Qaeda Threat a Priority Before 9/11." *The Washington Post*, 25 March 2004, A01.

REFERENCES (Continued)

Ekman, Paul. *Telling Lies: Clues to Deceit in the Marketplace, Politics, and Marriage.* New York, NY: W.W. Norton, 2009.

Emerson, Ralph Waldo. *The Essay on Self-Reliance.* East Aurora, NY: The Roycrafters, 1908.

Ennis, Robert. "A Taxonomy of Critical Thinking Skills and Dispositions." In Joan Boyloff Baron and Robert J. Sternberg, Eds. *Teaching Thinking Skills: Theory and Practice.* New York, NY: Freeman, 1987, 9-26.

Ericsson, K. Anders, Neil Charness, Robert R. Hoffman, and Paul J. Feltovich, eds. *Cambridge Handbook of Expertise and Expert Performance.* Cambridge, UK: Cambridge University Press, 2006.

Faculty of the School for Advanced Military Studies. *The Art of Design,* Volume 2. Fort Leavenworth, KS, 2010. URL: <http://www.cgsc.edu/events/sams/ArtofDesign_v2.pdf>. Accessed 20 May 2010.

Fagan, Garrett G. *Archaeological Fantasies: How Pseudoarchaeology Misrepresents the Past and Misleads the Public.* London, UK: Routledge, 2006.

Fallows, James. "Countdown to a Meltdown." *The Atlantic Monthly,* July/August 2005. URL: <http://www.theatlantic.com/doc/200507/fallows>. Accessed 15 February 2010.

Feltovich, Paul J., Robert R. Hoffman, Axel Roesler, David Woods. "Keeping It Too Simple: How the Reductive Tendency Affects Cognitive Engineering." *IEEE Intelligent Systems* 19, no. 3 (May/June 2004): 90-94.

Financial Times. "Executive Education — Open — 2009." URL: <http://rankings.ft.com/businessschoolrankings/executive-education---open>. Accessed 15 January 2010.

Fischoff, Baruch. "For Those Condemned to Study the Past: Heuristics and Biases in Hindsight." In Daniel Kahneman, Paul Slovic, and Amos Tversky. *Judgment Under Uncertainty: Heuristics and Biases.* Cambridge, UK: Cambridge University Press, 1982, 335-351.

Fishbein, Warren and Gregory Treverton. "Making Sense of Transnational Threats." *Sherman Kent Center Occasional Papers* 3, no. 1 (October 2004).

Fishman Barry J. and Samuel F. O'Connor-Divelbiss, eds. Fourth International Conference of the Learning Sciences. Mahwah, NJ: Erlbaum, 2000.

Folker, Robert D., Jr., MSgt, USAF. *Intelligence Analysis in Theater Joint Intelligence Centers: An Experiment in Applying Structured Methods*, Occasional Paper Number Seven. Washington, DC: Joint Military Intelligence College, 2000.

Ford, Harold P. "The CIA and Double Demonology: Calling the Sino-Soviet Split." *Studies in Intelligence* 42, no. 5 (Winter 1988-1989): 57-71.

Ford, Harold P. *CIA and Vietnam Policymakers: Three Episodes 1962-1968*. Washington, DC: Center for the Study of Intelligence, 1998.

Foreman-Wernet, Lois. "Rethinking Communication: Introducing the Sense-Making Methodology." In Brenda Dervin, Lois Foreman-Wernet, and Eric Lauterbach. *Sense-Making Methodology Reader: Selected Writings of Brenda Dervin*. Cresskill, NJ: Hampton Press, Inc., 2003, 1-10.

Fukuyama, Francis. "The Challenges of Uncertainty: An Introduction." In Francis Fukuyama, ed. *Blindside: How to Anticipate Forcing Events and Wild Cards in Global Politics*. Washington, DC: Brookings Institution Press, 2007, 1-6.

Fukuyama, Francis, ed. *Blindside: How to Anticipate Forcing Events and Wild Cards in Global Politics*. Washington, DC: Brookings Institution Press, 2007.

Garthoff, Raymond L. "US Intelligence in the Cuban Missile Crisis." In James G. Blight and David A. Welch, Eds. *Intelligence and the Cuban Missile Crisis*. London, UK: Frank Cass, 1998, 18-63.

Genomics and Genetics Weekly. "Why do People 'Play the Longshot' and Buy Insurance? It's in Our Genes." 29 January 2010: 184.

George, Roger Z. and James B. Bruce. *Analyzing Intelligence: Origins, Obstacles, and Innovations*. Washington, DC: Georgetown University Press, 2008.

REFERENCES (Continued)

Gibbons Robert and Andrew Rutten. "Institutional Interactions: An Equilibrium Approach to the State and Civil Society." *IQ* online journal. The Institute for Quantitative Social Science at Harvard University. URL: <http://www.iq.harvard.edu/files/iqss/old/PPE/gibbons+rutten.pdf>. Accessed 14 May 2010.

Gigerenzer, Gerd. *Simple Heuristics That Make Us Smart.* Oxford, UK: Oxford University Press, 1999.

Gilovich, Thomas, Dale Griffin, Daniel Kahneman. *Heuristics and Biases.* Cambridge, UK: Cambridge University Press, 2002.

Goldenberg, Jacob, David Mazursky, and Sorin Solomon. "Essays on Science and Society: Creative Sparks." *Science* 285, no. 5433 (3 September 1999): 1495-1496.

Gonzales, Laurence. *Deep Survival: Who Lives, Who Dies, and Why.* New York, NY: W.W. Norton, 2003.

Gorfein David S. and Robert Hoffman, eds. *Memory and Learning: The Ebbinghaus Centennial Conference.* Hillsdale, NJ: Erlbaum, 1982.

Gould, Stephen J. *The Mismeasure of Man.* New York, W. W. Norton & Company, 1996.

Greenberger, Martin, ed. *Computers, Communications, and the Public Interest.* Baltimore, MD: Johns Hopkins University Press, 1971.

Gross, Samuel R., Kristen Jacoby, Daniel J. Matheson, Nicholas Montgomery, and Sujata Patil. "Exonerations in the United States (1989 through 2003)." *Journal of Criminal Law and Criminology* 95, no. 2 (2005): 523-560.

Grove, William M. "Clinical Versus Statistical Prediction: The Contribution of Paul E. Meehl." *Journal of Clinical Psychology* 61, no. 10 (2005): 1233-1243.

Grove, William M., David H. Zald, Boyd S. Lebow, Beth E. Snitz, and Chad Nelson. "Clinical Versus Mechanical Prediction: A Meta-Analysis," *Psychological Assessment* 12, no. 1 (2000): 19-30.

Haines Gerald K. and Robert E. Leggett, eds. *Watching The Bear: Essays on CIA's Analysis of the Soviet Union.* Washington, DC: Center for the Study of Intelligence, 2003.

REFERENCES (Continued)

Hampson, Fen Osler. "The Divided Decision-Maker: American Domestic Politics and the Cuban Crises." *International Security* 9, no. 3 (Winter 1984-1985): 130-165.

Hansen, James H. "Soviet Deception in the Cuban Missile Crisis." *Studies in Intelligence* 46, no. 1 (2002): 49-58.

Hasher, Lynn, David Goldstein, and Thomas Toppino. "Frequency and the Conference of Referential Validity." *Journal of Verbal Learning and Verbal Behavior* 16 (1977): 107-112.

Hayden, GEN Michael V. (Ret.). Interview on national security and U.S. strategy. C-SPAN Radio. 9 August 2010. URL: <http://www.cspan. org/Watch/Media/2010/08/09/Terr/A/36779/Gen+Michael+ Hayden+Ret+The+Chertoff+Group+Principal.aspx>. Accessed 11 August 2010.

Hayek, Friedrich August. *The Counter-Revolution of Science: Studies on the Abuse of Reason*, 2nd Edition. Indianapolis, Indiana: Liberty Fund, Inc., 1979.

Herlihy, David. *The Black Death and the Transformation of the West.* Samuel K. Cohn, Jr., ed. Cambridge, MA: Harvard University Press, 1995.

Herrmann Douglas J. and Roger Chaffin, R. "Memory before Ebbinghaus." In David S. Gorfein and Robert Hoffman, eds. *Memory and Learning: The Ebbinghaus Centennial Conference.* Hillsdale, NJ: Erlbaum, 1982.

Hernstein, Richard J. and Charles Murray. *The Bell Curve: Intelligence and Class Structure in American Life.* New York, NY: The Free Press, 1996.

Hertwig, Ralph, Gerd Gigerenzer, and Ulrich Hoffrage. "The Reiteration Effect in Hindsight Bias." *Psychological Review* 104, no. 1 (1997): 194-202.

Heuer, Richards J., Jr. *Psychology of Intelligence Analysis.* Washington, DC: Center for the Study of Intelligence, 1999.

Heuer, Richards J., Jr. and Randolph H. Pherson. *Structured Analytic Techniques for Intelligence.* Washington, DC: CQ Press, 2010.

REFERENCES (Continued)

Hilton, Denis. "Causality vs. Explanation: Objective Relations vs. Subjective Interests." *Interdisciplines*. Institute of Cognitive Sciences. University of Geneva. URL: <http://www.interdisciplines.org/causality/papers/14>, accessed 1 November 2010.

Hoc, Jean-Michel, Pietro C. Cacciabue, and Erik Hollnagel, eds. *Expertise and Technology: Cognition and Human-Computer Cooperation*. Hillsdale, NJ: Lawrence Erlbaum Associates, 1995.

Hoffman, Robert. R. "How Can Expertise Be Defined? Implications of Research from Cognitive Psychology." In Robin Williams, Wendy Faulkner and James Fleck, eds. *Exploring Expertise*. New York, NY: MacMillan, 1998, 81-100.

Hoffman, Robert. R. "Biased about Biases: The Theory of the Handicapped Mind in *The Psychology of Intelligence Analysis*." Panel Presentation for "Designing Support for Intelligence Analysts," S. Potter, Chair. In *Proceedings of the 48th Annual Meeting of the Human Factors and Ergonomics Society*. Santa Monica, CA: Human Factors and Ergonomics Society, 2005, 406-410.

Hoffman, Robert. Conversation with the author, 4 October 2007.

Hoffman Robert R. and Laura Grace Militello. *Perspectives on Cognitive Task Analysis: Historical Origins and Modern Communities of Practice*. Boca Raton, FL: CRC Press/Taylor and Francis, 2008.

Holden, Constance. "Parsing the Genetics of Behavior." *Science* 322, no. 5903 (7 November 2008): 892-895.

Horn, Robert. Conversation with the author, 6 October 2010

IARPA. *Integrated Cognitive-Neuroscience Architectures for Understanding Sensemaking*. Broad Area Announcement IARPA-BAA-10-04, 1 April 2010. URL: <http://www.iarpa.gov/solicitations_icarus.html>. Accessed 1 June 2010.

James, William. *Principles of Psychology,* two volumes. New York, NY: Henry Holt and Company, 1890.

Jervis, Robert. "Why Intelligence and Policymakers Clash." *Political Science Quarterly* 125, no. 2 (Summer 2010): 185-204.

REFERENCES (Continued)

Johnston, Rob. *Analytic Culture in the U.S. Intelligence Community: An Ethnographic Study*. Washington, DC: Center for the Study of Intelligence, 2005.

Jones, Morgan, D. *The Thinker's Toolkit: 14 Powerful Techniques for Problem Solving*, Revised and Updated. New York, NY: Crown Publishing: 1998.

Kahneman Daniel, and Gary Klein. "Conditions for Intuitive Expertise: A Failure to Disagree." *American Psychologist* 64, no. 6 (September 2009): 515-526.

Kahneman, Daniel and Amos Tversky. "Prospect Theory: An Analysis of Decision Under Risk." *Econometrica* 47, no. 2 (March 1979): 263-291.

Kahneman, Daniel, Paul Slovic, and Amos Tversky. *Judgment Under Uncertainty: Heuristics and Biases*. Cambridge, UK: Cambridge University Press, 1982.

Kaiser Family Foundation. "WHO Rejects Accusations It Mishandled H1N1, Updates Worldwide Stats." Kaiser Family Foundation. URL: <http://globalhealth.kff.org/Daily-Reports/2010/January/25/GH-012510-Swine-Flu.aspx?utm_source=feedburner&utm_medium=feed&utm_campaign=Feed%3A+kff%2Fkdghpr+%28Kaiser+Daily+Global+Health+Policy+Report%29>. Accessed 2 February 2010.

Kendall, Willmoore. "The Function of Intelligence." *World Politics* 1, no. 4 (July 1949): 542-552.

Kent, Sherman. "A Crucial Estimate Relived." *Studies in Intelligence* 8, no. 2 (Spring 1964): 1-18 (originally classified SECRET). Declassified and reprinted in vol. 36, no. 5 (1992): 111-119.

Kent, Sherman. *Strategic Intelligence for American World Policy*. Princeton, NJ: Princeton University Press, 1949.

Kerbel, Josh and Anthony Olcott. "Synthesizing with Clients, Not Analyzing for Customers." *Studies in Intelligence* 54, no. 4 (December 2010): 1-13.

Kilbourne, Edwin D. "Influenza Pandemics of the 20th Century." *Emerging Infectious Diseases* 12, no. 1 (January 2006): 9-14. URL: <www.cdc.gov/eid>. Accessed 30 March 2010.

REFERENCES (Continued)

Klein, Gary and Robert R. Hoffman, "Macrocognition, Mental Models, and Cognitive Task Analysis Methodology." In Jan Maarten Schraagen, Laura Grace Militello, Tom Ormerod, and Raanan Lipshitz, eds. *Naturalistic Decision Making and Macrocognition.* Aldershot, UK: Ashgate Publishing Limited, 2008, 57-80.

Klein, Gary. "Flexecution as a Paradigm for Replanning, Part 1." *IEEE Intelligent Systems* 22, no. 5 (September/October 2007): 79-88.

Klein, Gary. "Flexecution as a Paradigm for Replanning, Part 2." *IEEE Intelligent Systems* 22, no. 6 (November/December 2007): 108-112.

Klein, Gary, Brian Moon, and Robert R. Hoffman. "Making Sense of Sensemaking 1: Alternative Perspectives." *IEEE Intelligent Systems* 21, no. 4 (July/August 2006): 70-73.

Klein, Gary, Brian Moon, and Robert R. Hoffman. "Making Sense of Sensemaking 2: A Macrocognitive Model." *IEEE Intelligent Systems* 21, no. 5 (September/October 2006): 88-92.

Klein, Gary, Karol G. Ross, Brian M. Moon, Devorah E. Klein, Robert R. Hoffman, and Erik Hollnagel. "Macrocognition." *IEEE Intelligent Systems* 18, no. 3 (May/June 2003): 81-85.

Kleinmuntz, Benjamin. "The Processing of Clinical Information by Man and Machine." In Benjamin Kleinmuntz, ed. *Formal Representations of Human Judgment.* New York, NY: Wiley, 1968, 149-186.

Knowles, David. *Great Historical Enterprises and Problems in Monastic History.* London, UK: Thomas Nelson and Sons, 1962.

Kosso, Peter. "Introduction: The Epistemology of Archaeology." In Garrett G. Fagan, *Archaeological Fantasies: How Pseudoarchaeology Misrepresents the Past and Misleads the Public.* London, UK: Routledge, 2006, 3-22.

Krizan, Lisa A. *Intelligence Essentials for Everyone.* Occasional Paper Number Six. Washington, DC: Joint Military Intelligence College, 1999.

Kuhn, Thomas S. *The Structure of Scientific Revolutions.* Chicago, IL: University of Chicago Press, 1962.

REFERENCES (Continued)

Kuhn, Thomas S. *The Road Since Structure: Philosophical Essays, 1970-1993, with an Autobiographical Interview*. James Conant and John Haugeland, eds. Chicago, IL: University of Chicago Press, 2000.

Langer, Ellen J. *Mindfulness*. Cambridge, MA: Da Capo Press, 1989.

Leedom, Dennis K. *Final Report Sensemaking Symposium*. Command and Control Research Program Office of the Assistant Secretary of Defense for Command, Control, Communications and Intelligence, 23-25 October 2001.

Lester, Gregory W. "Why Bad Beliefs Don't Die." *Skeptical Inquirer* 24, no. 6 (November/December 2000). URL: <http://www.csicop.org/si/archive/category/volume_24.6>. Accessed 22 September 2009.

Lin, Lin. "Breadth-Biased Versus Focused Cognitive Control in Media Multitasking Behaviors." *Proceedings of the National Academies of Science* 106, no. 37 (15 September 2009): 15521-15522. URL: <www.pnas.org!cgi!doi!10.1073!pnas.0908642106 PNAS>. Accessed 18 March 2010.

MacEachin, Douglas. *Predicting the Soviet Invasion of Afghanistan: The Intelligence Community's Record*. Washington, DC: Center for the Study of Intelligence, 2002.

Macintyre, Ben. *Operation Mincemeat*. New York, NY: Harmony Books, 2010.

MacLean, Katherine A., Emilio Ferrer, Stephen R. Aichele, David A. Bridwell, Anthony P. Zanesco, Tonya L. Jacobs, Brandon G. King, Erika L. Rosenberg, Baljinder K. Sahdra, Phillip R. Shaver, B. Alan Wallace, George R. Mangun, and Clifford D. Saron. "Intensive Meditation Training Improves Perceptual Discrimination and Sustained Attention." *Psychological Science* 21, no. 6 (2010): 829-830.

Mandel, David R., Alan Barnes, and John Hannigan. "A Calibration Study of an Intelligence Assessment Division." PowerPoint Presentation. Defense Research and Development Canada, Toronto, CA, not dated.

Manjoo, Farhad. *True Enough: Learning to Live in a Post-Fact Society*. Hoboken, NJ: John Wiley & Sons, 2008.

REFERENCES (Continued)

Mccosh, James. LL.D. *Intuitions of the Mind: Inductively Investigated.* London: UK: Macmillan and Company, 1882.

McCallister, William S. "COIN and Irregular Warfare in a Tribal Society." *Small Wars Journal*, blog and pamphlet, 4 February 2008. URL: <http://www.smallwarsjournal.com/documents/coinandiwinatribalsociety.pdf>. Accessed 31 May 2010.

Menkhaus, Kenneth. "State Fragility as a Wicked Problem," *Prism* 1, no. 2 (March 2010): 85-100. URL: <http//:www.ndu.edu/press/lib/images/prism1-2/6_Prism_85-100_Menkhaus.pdf>. Accessed 14 May 2010.

Meszerics, Tamas and Levente Littvay, "Pseudo-Wisdom and Intelligence Failures," *International Journal of Intelligence and Counterintelligence* 23, no. 1 (December 2009): 133-147.

Mercyhurst College Institute for Intelligence Studies. "Non-State Actors in Sub-Saharan Africa 2007-2012 Outlook." URL: <https://nonstateactorsafrica.wikispaces.com>. Accessed 28 April 2010.

Miller, George A. "The Magical Number Seven, Plus or Minus Two: Some Limits on Our Capacity for Processing Information." *The Psychological Review* 63 (1956): 81-97.

Miller, Linda K. and Mary McAuliffe. "The Cuban Missile Crisis," *Magazine of History* 8 (Winter 1994). URL: <http://www.oah.org/pubs/magazine/coldwar/miller.html>. Accessed 5 January 2010.

Miller, Peter. *The Smart Swarm: How Understanding Flocks, Schools, and Colonies Can Make Us Better at Communicating, Decision Making, and Getting Things Done.* New York, NY: Penguin Books, 2010.

Monk Christopher A., J. Gregory Trafton, and Deborah A. Boehm-Davis. "The Effect of Interruption Duration and Demand on Resuming Suspended Goals." *Journal of Experimental Psychology: Applied* 14 (December 2008): 299-313.

Montagu, Ewen. *The Man Who Never Was: World War II's Boldest Counterintelligence Operation.* Annapolis, MD: Naval Institute Press, 2001.

Moore, David T. *Creating Intelligence: Evidence and Inference in the Analysis Process.* MSSI Thesis chaired by Francis J. Hughes. Washington, DC: Joint Military Intelligence College, July 2002.

REFERENCES (Continued)

Moore, David T. *Critical Thinking and Intelligence Analysis*, Occasional Paper Number Fourteen, Revised. Washington, DC: National Defense Intelligence College, 2009.

Moore, David T., and Lisa Krizan. "Core Competencies for Intelligence Analysis at the National Security Agency." In Russell G. Swenson, ed., *Bringing Intelligence About: Practitioners Reflect on Best Practices*. Washington, DC: Joint Intelligence Military College, 2003, 95-132.

Moore, David T., Lisa Krizan, and Elizabeth J. Moore. "Evaluating Intelligence: A Competency-Based Approach." *International Journal of Intelligence and CounterIntelligence* 18, no. 2 (Summer 2005): 204-220.

Moore, David T. and William N. Reynolds, "So Many Ways To Lie: The Complexity of Denial and Deception." *Defense Intelligence Journal* 15, no. 2 (Fall 2006): 95-116.

Morens, David M. and Jeffery K. Taubenberger. "Influenza and the Origins of The Phillips Collection, Washington, DC." *Emerging Infectious Diseases* 12, No. 1 (January 2006): 78-80. URL: <www.cdc.gov/eid>, accessed 30 March 2010.

Mosier, Kathleen and Ute Fischer, eds. *Proceedings of the Eighth International NDM Conference*. Eighth International Naturalistic Decision-Making Conference, Pacific Grove, CA, June 2007.

Myers, David G. *Intuition: Its Powers and Perils*. New Haven, CT: Yale University Press, 2002.

Mylroie, Laurie. "Who is Ramzi Yousef? And Why It Matters." *The National Interest*, 22 December 1995. URL: <http://www.fas.org/irp/world/iraq/956-tni.htm>. Accessed 12 October 2010.

National Commission on Terrorist Attacks Upon the United States. *The 9/11 Commission Report*. Washington, DC: Government Printing Office, 2004.

National Intelligence Council. *Global Trends 2015: A Dialogue About the Future with Nongovernmental Experts*. Washington, DC: Government Printing Office, 2000. URL: <http://www.dni.gov/nic/NIC_globaltrend2015.html>. Accessed 15 February 2010.

REFERENCES (Continued)

National Intelligence Council. "Nonstate Actors: Impact on International Relations and Implications for the United States." Conference Report, August 2007. URL: <http://www.dni.gov/nic/confreports_nonstate_actors.html>. Accessed 27 April 2010.

National Intelligence Council. *Global Trends 2025: A Transformed World.* Washington, DC: Government Printing Office. URL: <http://www.odni.gov/>. Accessed 12 December 2008.

Neustadt, Richard E. and Ernest R. May. *Thinking in Time: The Uses of History for Decisionmakers.* New York, NY: The Free Press, 1986.

New Oxford American Dictionary, Apple Computer Edition, 2005. Various entries.

Nolte, William. "Keeping Pace with the Revolution in Military Affairs." *Studies in Intelligence*, 48, no. 1 (Winter 2004): 1-10.

Nosich, Gerald R. *Learning to Think Things Through: A Guide to Critical Thinking Across the Curriculum*, 3rd edition. Upper Saddle River, NJ: Prentice Hall, 2008.

Obama, Barack. "Remarks by the President on Security Reviews," 5 January 2010. URL: <http://www.whitehouse.gov/the-press-office/remarks-president-security-reviews>. Accessed 7 January 2010.

Oberdorfer, Don. "Missed Signals in the Middle East, *The Washington Post Magazine*, 17 March 1991: 19-41.

Office of the Director of National Intelligence. *Intelligence Community Directive Number 203: Analytic Standards* (21 June 2007). URL: <http://www.fas.org/irp/dni/icd/icd-203.pdf>. Accessed 16 October 2010.

Olcott, Anthony. "Revisiting the Legacy: Sherman Kent, Willmoore Kendall, and George Pettee — Strategic Intelligence in the Digital Age." *Studies in Intelligence* 51, no. 2 (June 2009): 21-32.

Ophir, Eyal, Clifford Nass, and Anthony D. Wagner. "Cognitive Control in Media Multitaskers." *Proceedings of the National Academies of Science, Early Edition* 106, no. 37: 15583-15587. URL: < http://www.pnas.org/content/106/37/15583.full>. Accessed 18 March 2010.

REFERENCES (Continued)

Pappas, Aris A., and James M. Simon, Jr. "The Intelligence Community: 2001-2015: Daunting Challenges, Hard Decisions," *Studies in Intelligence* 46, no. 1 (2002). URL: <http://www.cia.gov/csi/studies/vol46no1/article05.html>. Accessed 10 January 2006.

Park, Caroline L. "What Is the Value of Replicating other Studies?" *Research Evaluation* 13. no. 1 (December 2004): 189-195.

Patterson Emily S. and Janet E. Miller. *Macrocognition Metrics and Scenarios: Design and Evaluation for Real World Teams.* Aldershot, UK: Ashgate, 2010.

Paul, Richard W. and Linda Elder. *Critical Thinking: Tools for Taking Charge of Your Professional and Personal Life.* Upper Saddle River, NJ: Prentice Hall, 2002.

Paul, Richard and Linda Elder. *The Miniature Guide to Critical Thinking: Concepts and Tools*, Sixth edition. Dillon Beach, CA: Foundation for Critical Thinking, 2009.

Peirce, Charles Sanders. "Lowell Lectures (1903)." In Charles Hartshorne and Paul Weiss, eds. *Collected Papers of Charles Sanders Peirce, Volume 1.* Cambridge, MA: Harvard University Press, 1931.

Peirce, Charles Sanders. "A Letter to Lady Welby (1904)." In Arthur W. Burks, ed. *Collected Papers of Charles Sanders Peirce, Volume 8.* Cambridge, MA: Harvard University Press, 1958.

Pezzo, Mark V. and Stephanie P. Pezzo, "Making Sense of Failure: A Motivated Model of Hindsight Bias," *Social Cognition* 25, no. 1 (2007): 147-164.

Phelps, R. H. and James Shanteau, "Livestock Judges: How Much Information Can an Expert Use?" *Organizational Behavior and Human Performance* 21 (1978): 213-222.

Pirolli, Peter. *Information Foraging Theory: Adaptive Interaction with Information.* Oxford, UK: Oxford University Press, 2007.

Pirolli, Peter, and Stuart Card. "The Sensemaking Process and Leverage Points for Analyst Technology as Identified Through Cognitive Task Analysis." Paper presented a the 2005 International Conference on Intelligence Analysis, Vienna, Virginia, 2-6 May 2005, URL: <https://analysis.mitre.org/proceedings_agenda.htm#papers>. Accessed 11 March 2009.

REFERENCES (Continued)

Pirolli, Peter and Stuart Card, "Information Foraging," *Psychological Review* 106, no. 4 (October 1999): 643-675.

Pitt, Will. "Interview: 27-Year CIA Veteran." *Truthout*, 26 June 2003. URL: <http://www.truthout.org/docs_03/ 062603B.shtml>. Accessed 12 March 2007.

Planck, Max. *Scientific Autobiography and other Papers*, F. Glynor, trans. New York, NY: Philosophical Library, 1949.

Raeburn, Paul. "Multitasking May Not Mean Higher Productivity." NPR News: *Science Friday*, 28 August 2009. URL: <http://www.npr.org/ templates/story/story.php?storyId=112334449&ft=1&f=1007>, accessed 21 February 2010.

Rand, Ayn. *Introduction to Objectivist Epistemology*. New York, NY: Mentor, 1979.

Reed, Jennifer H. "Effect of a Model For Critical Thinking on Student Achievement In Primary Source Document Analysis And Interpretation, Argumentative Reasoning, Critical Thinking Dispositions, and History Content in a Community College History Course." PhD Dissertation. College of Education, University of South Florida, December 1998. URL: <http:// www.criticalthinking.org/resources/JReed-Dissertation.pdf>. Last accessed 2 February 2010.

Reed, Nick, and R. Robbins. "The Effect of Text Messaging on Driver Behaviour." *Published Project Report PPR 367*. RAC Foundation Transport Research Laboratory, September 2008. URL: <http:// www.racfoundation.org/files/textingwhiledrivingreport.pdf>. Accessed 9 December 2009.

Reynolds, William N. and others. "Social Science Modeling Workshop: Understanding Iraqi Non-State Actors." *Workshop Proceedings*. Least Squares Software, Albuquerque, NM, 15 February 2008.

Richardson, Kurt A., ed. *Managing Organizational Complexity: Philosophy, Theory, and Application*. Greenwich, CT: Information Age Publishing, 2005: 7-19.

Ricks, Thomas E. "A Marine's Afghan AAR (XIV): Get Rid of the iPods on Patrol," Web Log, *Foreign Policy.* Weblog Entry, 20 January 2010. URL: <http://ricks.foreignpolicy.com/posts/2010/01/20/a_marine_s_afghan_aar_xiv_get_rid_of_ the_ipods_on_patrol>. Accessed 26 January 2009.

Rieber, Steven. "Intelligence Analysis and Judgmental Calibration." *International Journal of Intelligence and CounterIntellgience* 17, no. 1 (Spring 2004): 97-112.

Rieber, Steven, and Neal Thomason. "Towards Improving Intelligence Analysis: Creation of a National Institute for Analytic Methods." *Studies in Intelligence.* Online edition 49, no. 4 (Fall 2005). URL: <https://www.cia.gov/library/center-for-the-study-of-intelligence/csi-publications/csi-studies/studies/vol49no4/Analytic_Methods_7.htm#_ftn10>. Accessed 7 January 2009.

Rihoux, Benoit. "Constructing Political Science Methodology: From Segmented Polarization to Enlightened Pluralism." Joint Chair, Standing Group on Political Methodology. *IPSA Conference.* Montreal, Canada, 2008.

Rittel, Horst W. J. and Melvin M. Webber. "Dilemmas in a General Theory of Planning." *Policy Sciences* 4 (1973): 155-169.

Robb, John. *Brave New War: The Next Stage of Terrorism and the End of Globalization.* New York, NY: John A. Wiley and Sons, 2007.

Rochman, Bonnie. "Samurai Mind Training for Modern American Warriors." *Time.* Online edition, 6 September 2009. URL: <http://www.time.com/time/nation/article/0,8599,1920753,00.html>. Accessed 4 August 2010.

Rogoff, Barbara and Jean Lave. *Everyday Cognition: Its Development in Social Context.* Cambridge, MA: Harvard University Press, 1984.

Ronson, Jon. "What Part of 'Bomb' Don't You Understand?" *This American Life,* Episode 338, 3 August 2008. URL: <http://www.thislife.org/Radio_Episode.aspx?episode=338>. Accessed 4 August 2008.

Rosenhead, Jonathan, ed. *Rational Analysis for a Problematic World: Problem Structuring Methods for Complexity, Uncertainty and Conflict,* 2nd edition. Hoboken, NJ: John Wiley & Sons, 2001.

REFERENCES (Continued)

Rosenhead, Jonathan. "Complexity Theory and Management Practice." URL: <http://www.human-nature.com/science-as-culture/rosenhead.html. Accessed 23 December 2008.

Rummel, Rudolph. J. *Understanding Conflict and War: War, Power, Peace.* Beverly Hills, CA: Sage Publications, 1991.

Russell, Dennis M., Mark J. Stefik, Peter Pirolli, and Stuart Card. "The Cost Structure of Sensemaking." Paper presented a the INTERCHI '93 Conference on Human Factors in Computing Systems, Amsterdam, NL, 24-25 April 1993. URL: <http://www2.parc.com/istl/groups/uir/publications/items/UIR-1993-10-Russell.pdf>. Accessed 18 August 2010.

Schacter, Daniel L., Donna Rose Addis and Randy L, Buckner. "Remembering the Past to Imagine the Future: The Prospective Brain," *Nature Review Neuroscience* 8 (September 2007): 657-661.

Schraagen, Jan, Maarten, Laura Grace Militello, Tom Ormerod, and Raanan Lipshitz, eds. *Naturalistic Decision Making and Macrocognition.* Aldershot, UK: Ashgate Publishing Limited, 2008.

Schreiber Darren M., Alan N. Simmons,, Christopher T. Dawes, Taru Flagan, James H. Fowler, and Martin P. Paulus. "Red Brain, Blue Brain: Evaluative Processes Differ in Democrats and Republicans." Paper delivered at the 2009 American Political Science Association Meeting, Toronto, CA, 3-6 September 2009. URL: <http ://ssrn. com/abstract=1451867>. Accessed 9 December 2009.

Schrodt Philip A. and Deborah J. Gerner. "The Impact of Early Warning on Institutional Responses to Complex Humanitarian Crises." Paper presented at the Third Pan-European International Relations Conference and Joint Meeting with the International Studies Association. Vienna, Austria, 16-19 September 1998.

Schum, David A. *Evidence and Inference for the Intelligence Analyst,* two volumes. Lanham, MD: University Press of America, 1987.

Schwartz, Peter. *The Art of the Long View.* New York, NY: Doubleday, 1996.

Schwartz, Peter, and Doug Randall. "Ahead of the Curve: Anticipating Strategic Surprise." In Francis Fukuyama, ed. *Blindside: How to Anticipate Forcing Events and Wild Cards in Global Politics.* Washington, DC: Brookings Institution Press, 2007: 93-108.

REFERENCES (Continued)

Scribner, Sylvia. "Studying Working Intelligence. Everyday Cognition: Its Development in Social Context." In Barbara Rogoff and Jean Lave. *Everyday Cognition: Its Development in Social Context.* Cambridge, MA: Harvard University Press, 1984, 9-40.

Scribner, Sylvia. "Thinking in Action: Some Characteristics of Practical Thought." In Ethel Tobach, Rachel Joffe Falmagne, Mary Brown Parlee, Laura M. W. Martin, and Aggie Scribner Kapelman, eds. *Mind and Social Practice: Selected Writings of Sylvia Scribner.* Cambridge, UK: Cambridge University Press, 319-337.

Sethi, Deepak. "Mindful Leadership." *Leader to Leader* 51 (Winter 2009): 7-11.

Shanteau, James. "Competence in Experts: The Role of Task Characteristics." *Organizational Behavior and Human Decision Processes* 53 (1992): 252-266.

Shryock, Richard. "For An Eclectic Sovietology." *Studies in Intelligence* 8, no. 1 (Winter 1964): 57-64.

Simon, Herbert A. "Designing Organizations in an Information-Rich World." In Martin Greenberger, ed. *Computers, Communications, and the Public Interest.* Baltimore, MD: Johns Hopkins University Press, 1971, 40-41.

Slovic, Paul. "Behavioral Problems of Adhering to a Decision Policy." Paper presented at the Institute for Quantitative Research in Finance, Napa, CA, May 1973.

Smith, Rupert. *The Utility of Force: The Art of War in the Modern World.* New York, NY: Alfred A. Knopf, 2007.

Snowden, David, PhD. Conversation with the author, 22 January 2008.

Spelke, Elizabeth S., William Hirst, and Ulric Neisser. "Skills of Divided Attention." *Cognition* 4 (1976): 215-230.

Stanger-Hall, Kathrin. "Accommodation or Prediction?" Letter in response to Peter Lipton, "Testing Hypotheses: Prediction and Prejudice." *Science* 308 (3 June 2005): 1409.

Stefik, Mark. "The New Sensemakers: The Next Thing beyond Search is Sensemaking." *Innovation Pipeline* (15 October 2004). URL: <http://www.parc.com/research/publications/files/5367.pdf>. Accessed 11 March 2009.

REFERENCES (Continued)

Stephens, David W., and John R. Krebs. *Foraging Theory.* Princeton, NJ: Princeton University Press, 1986.

Steury, Donald P., Ed. *Sherman Kent and the Board of National Estimates,* Collected Essays. Washington, DC: Center for the Study of Intelligence, 1994.

Swenson, Russell G., ed. *Bringing Intelligence About: Practitioners Reflect on Best Practices.* Washington, DC: Joint Military Intelligence College, 2003.

Taleb, Nassim Nicholas. *The Black Swan: The Impact of the Highly Improbable.* New York, NY: Random House, 2007.

Taleb, Nassim Nicholas. *Fooled by Randomness: The Hidden Role of Chance in Life and in the Markets,* 2nd revised edition. New York, NY: Random House, Inc., 2008.

Tavris, Carol and Elliot Aronson. *Mistakes Were Made (But Not by Me): Why We Justify Foolish Beliefs, Bad Decisions, and Hurtful Acts.* Orlando, FL: Harcourt, 2007.

Thomas, Troy S., Stephen D. Kiser, and William D. Casebeer. *Warlords Rising: Confronting Violent Non-State Actors.* Lanham, MD: Lexington Books, 2005.

Tobach, Ethel, Rachel Joffe Falmagne, Mary Brown Parlee, Laura M. W. Martin, and Aggie Scribner Kapelman, eds. *Mind and Social Practice: Selected Writings of Sylvia Scribner.* Cambridge, UK: Cambridge University Press, 1997.

Transportation Research Board of the National Academies. "Selected References on Distracted Driving: 2005-2009." URL: <http://pubsindex.trb.org/DOCs/Publications from TRIS on Distracted Driving.pdf>. Accessed 9 December 2009.

Treverton, Gregory F. *Intelligence for an Age of Terror.* Cambridge, UK: Cambridge University Press, 2009.

Treverton, Gregory F. *Reshaping National Intelligence for an Age of Information.* UK: Cambridge University Press, 2001.

Treverton, Gregory F. "Estimating Beyond the Cold War." *Defense Intelligence Journal* 3, no. 2 (Fall 1994): 5-20.

Treverton, Gregory F. and C. Bryan Gabbard. *Assessing the Tradecraft of Intelligence Analysis*. TR-293. Santa Monica, CA: RAND Corporation, 2008.

Tsouras, Peter G. *Disaster at D-Day: The Germans Defeat the Allies, June 1944*. London, UK: Greenhill Books, 2000.

Tsouras, Peter G. *Third Reich Victorious: Ten Dynamic Scenarios in Which Hitler Wins the War*. London, UK: Lionel Leventhal Limited, 2002.

Tversky, Amos and Daniel Kahneman. "Availability: A Heuristic for Judging Frequency and Probability." *Cognitive Psychology* 5 (1973): 207-232.

Tversky, Amos and Daniel Kahneman. "Advances In Prospect Theory: Cumulative Representation Of Uncertainty." *Journal of Risk and Uncertainty* 5 (1992): 297-323.

United States Commission on National Security. *New World Coming: American Security in the 21st Century*. Study Addendum, Phase 1 (July 1998-August 1999).

United States Office of Personnel Management. Workforce Compensation and Performance Service. *Introduction to the Position Classification Standards*, TS-107 August 1991. URL: <http://www.opm.gov/fedclass/gshbkocc.pdf>. Accessed 11 December 2009.

United States Office of Personnel Management. *Position Classification Standard for Intelligence Series*, GS-0132 TS-28 June 1960, TS-27 April 1960. URL: <http://www.opm.gov/fedclass/html/gsseries.asp>. Accessed 11 December 2009.

United States Senate. *Report on the U.S. Intelligence Community's Prewar Intelligence Assessments on Iraq*. Select Senate Committee on Intelligence. 108th Congress. 7 July 2004.

University of Bristol, "Adult Stem Cell Breakthrough: First Tissue-engineered Trachea Successfully Transplanted." *Science Daily*, 18 November 2008. URL: <http://www.sciencedaily.com/releases/2008/11/081119092939.htm>. Accessed 21 November 2008.

REFERENCES (Continued)

Van Gelder, Tim. "Mindfulness Versus Metacognition, and Critical Thinking." *Bringing Visual Clarity to Complex Issues*. Weblog Entry, 27 May 2009. URL: <http://timvangelder.com/2009/05/27/mindfulness-versus-metacognition-and-critical-thinking/>. Accessed 13 January 2010.

von Clausewitz, Carl. *On War*. COL James J. Graham, trans. London, UK: Keegan Paul, Trench, Truebner & Co., Ltd.,1908.

Wasserman, David, Richard O. Lempert, Reid Hastie. "Hindsight and Causality." *Personality and Social Psychology Bulletin*, vol. 17, no. 1 (February 1991): 30-35.

Weber, Marta S. Conversation with the author, 2 August 2009.

Weber, Marta S., William N. Reynolds, James Holden-Rhodes, and Elizabeth J. Moore. *Non-State Actors in the Post-Westphalian World Order: A Preliminary LSS Inquiry*. Final Report for Air Force Research Laboratory (AFRL) Contract FA8750-07-C-0312. March 2007.

Weick Karl E. and Kathleen M. Sutcliffe. *Managing the Unexpected: Assuring High Performance in an Age of Complexity*. San Francisco, CA: Jossey-Bass, 2001.

Weick, Karl E. *Sensemaking in Organizations*. Thousand Oaks, CA: Sage Publications, Inc., 1995.

Weiss, Charles. "Communicating Uncertainty in Intelligence and Other Professions." *International Journal of Intelligence and CounterIntelligence* 21, no. 1 (Spring 2008): 57-85.

Westrum, Ronald. "Social intelligence about hidden events." *Knowledge: Creation, Diffusion, Utilization* 3, no. 3 (1982): 381-400.

Wheaton, Kristan J. Email to William N. Reynolds, 15 November 2007.

Wheaton, Kristan J. Email to David T. Moore, 4 March 2008.

Wheaton, Kristan J. and others. *Structured Analysis of Competing Hypotheses*. Erie, PA: Mercyhurst College, 2005.

Wigmore, John Henry. *The Science of Proof: As Given by Logic, Psychology and General Experience and Illustrated in Judicial Trials*, 3rd edition. Boston, MA: Little, Brown, 1937.

REFERENCES (Continued)

Wikipedia, entry under "Analysis." URL: <http://en.wikipedia.org/wiki/Analysis>. Accessed 26 September 2007.

Wikipedia, entry under "Normal Science." URL: <http://en.wikipedia.org/wiki/Normal_science>. Accessed 26 September 2007.

Williams, Robin, Wendy Faulkner and James Fleck, eds. *Exploring Expertise.* New York, NY: MacMillan, 1998.

Winograd, Eugene and Ulric Neisser, Eds. *Affect and Accuracy in Recall: Studies of "Flashbulb" Memories.* Emory Symposia in Cognition. Cambridge, UK: Cambridge University Press, 1992.

Wohlstetter, Roberta. *Pearl Harbor: Warning and Decision.* Stanford, CA: Stanford University Press, 1962.

Woods, Kevin, James Lacey, and Williamson Murray. "Saddam's Delusions: The View From the Inside." *Foreign Affairs* (May/June 2006).

Young, Michael, Yi Guan, John Toman, Andy DePalma, Elena Znamenskaia. "Agent as Detector: An Ecological Psychology Perspective on Learning by Perceiving-Acting Systems." In Barry J. Fishman and Samuel F. O'Connor-Divelbiss, eds. *Fourth International Conference of the Learning Sciences.* Mahwah, NJ: Erlbaum, 2000, 299-301.

Zelik, Daniel J., Emily S. Patterson, and David D. Woods. "Understanding Rigor in Information Analysis." In Kathleen Mosier and Ute Fischer, eds. *Proceedings of the Eighth International NDM Conference.* Eighth International Naturalistic Decision-Making Conference, Pacific Grove, CA, June 2007, 1-8.

Zelik, Daniel J., Emily S. Patterson, and David D. Woods. "Measuring Attributes of Rigor in Information Analysis." In Emily S. Patterson and Janet E. Miller, and David D. Woods. *Macrocognition Metrics and Scenarios: Design and Evaluation for Real World Teams.* Aldershot, UK: Ashgate, 2010: 65-83.

Zelik, Daniel J. Email to the author, 15 October 2010.

Zukav, Gary. *The Dancing Wu Li Masters: An Overview of the New Physics.* New York, NY: Harper Collins, 2001.

Zwicky, Fritz. *Discovery, Invention, Research — Through the Morphological Approach.* Toronto, CA: The Macmillan Company, 1969.

ABOUT THE AUTHOR

David T. Moore is a senior intelligence professional and educator. He teaches critical thinking and structured techniques for intelligence sensemaking and his research focuses on developing multidisciplinary approaches to facilitate intelligence sensemaking. His most recent posting was to the School of Leadership and Professional Development at the National Geospatial-Intelligence Agency. Formerly he served as a technical director at the National Security Agency where he advocated and mentored best practices in intelligence sensemaking. He is an adjunct faculty member of the National Cryptologic School; has taught at Trinity University, Washington, DC; and lectures at the National Defense Intelligence College and the National Defense University. He received a Master of Science of Strategic Intelligence from the National Defense Intelligence College in 2002.

Mr. Moore's publications include:

- "Species of Competencies for Intelligence Analysis," *Defense Intelligence Journal* 11, no. 2 (Summer 2002): 97-119.

- "Species of Competencies for Intelligence Analysis," *American Intelligence Journal* 23 (2005): 29-43 (an expanded version of the original article).

- *Critical Thinking and Intelligence Analysis*, Occasional Paper Number Fourteen; Authorized, Revised Edition (Washington, DC: National Defense Intelligence College, 2006, 2009).

- With coauthor Lisa Krizan:
  - "Intelligence Analysis, Does NSA Have What It Takes?" *Cryptologic Quarterly* 20, nos. 1/2 (Summer/Fall 2001): 1-32.
  - "Core Competencies for Intelligence Analysis at the National Security Agency," in *Bringing Intelligence About: Practitioners Reflect on Best Practices*, Russell Swenson, ed. (2004): 95-131.

- With coauthors Lisa Krizan and Elizabeth J. Moore:
  - "Evaluating Intelligence: A Competency-Based Approach," *International Journal of Intelligence and CounterIntelligence* 18, no. 2 (Summer 2005): 204-220.

- With coauthor William N. Reynolds:
    - "So Many Ways To Lie: The Complexity of Denial and Deception," *Defense Intelligence Journal* 15, no. 2 (Fall 2006): 95-116.
    - "Advancing the Practice: Multi-methodological Analysis for Intelligence," forthcoming.
- With coauthor Robert R. Hoffman:
    - "Sensemaking: A Tranformative Activity," *American Intelligence Journal*, 29, no. 1 (Spring 2011): 26-36.
    - "Data-Frame Theory of Sensemaking as a Best Model for Intelligence," *American Intelligence Journal*, 29, no. 2 (Fall 2011): in press.
    - "How Might Critical Thinking and Structured Analytic Techniques Improve Intelligence? Or..." Paper presented at the 6th semiannual 5-Eyes Conference on the Teaching of Intelligence, Joint Base Anacostia-Bolling, Washington, DC, 1-3 November 2011.
- With coauthors Robert R. Hoffman, Simon Henderson, Brian Moon, and Jordan A. Litman:
    - Reasoning Difficulty in Analytical Activity," *Theoretical Issues in Ergonomics Science*, 12, no. 3 (May 2011): 225-240.

The author can be contacted via email at *david.t.moore@ugov.gov*.

ABOUT THE CONTRIBUTORS

Robert R. Hoffman, PhD. A world leader in the field of cognitive systems engineering and Human-Centered Computing, he is a Fellow of the Association for Psychological Science and a Fulbright Scholar. Following post-doctoral work at the Center for Research on Human Learning at the University of Minnesota, Dr. Hoffman joined the faculty of the Institute for Advanced Psychological Studies at Adelphi University. Dr. Hoffman has been recognized internationally for his research on cognitive task analysis and the design of macrocognitive work systems.

Elizabeth J. Moore is a senior intelligence professional. She holds a BA in Russian Studies from Randolph-Macon Woman's College, and an MA in International Politics from American University. Ms. Moore began her government career as a Russian linguist in 1982, and has served in a number of language and analysis positions at the Defense, State, and Treasury Departments; at the National Intelligence Council; as a Director of National Intelligence Fellow attached to the National Security Council; and overseas.

William N. Reynolds, PhD. The founder, President and Chief Science Officer of Least Squares Software, he has been a principal researcher and innovator in the field of complexity for more than twenty years. Over the past eight years, he has focused on the role of complex systems in intelligence analysis, especially complexity-based analytic methodologies. In this capacity, Dr. Reynolds has formulated numerous intelligence analysis case studies assessing the impact of technology on efficiency and effectiveness in analytic practice.

Marta S. Weber, PhD. A psychologist who has applied thirty years of clinical and forensic training and experience to the intelligence domain, Dr. Weber pioneered in the development of sophisticated remote profiling and behavioral forecasting techniques. She is an internationally recognized expert in that highly specialized field. She founded and heads Applied Behavioral Sciences, LLC, a boutique consultancy specializing in primary human-source, human-subject research, analysis, strategic consultation and training.

INDEX

A

Adversarial deception 19, 60, 85, 100
  Operation Mincemeat 85
Analysis
  definition xxxv, 16, 34
  function of ii, 16, 55, 150
  imprecision and inaccuracy of concept 43, 48
Analysis of competing hypotheses xi, 98, 100, 102, 119
  Folker's experiment 101
  Mercyhurst structured version 102
  MITRE critique 101
  value and limitations xi, 118-119
Analytic pathologies 25, 26
Anchoring 24, 119
  definition 79
Automatic thinking 70-72
  definition 70
  risks 71

C

Cognitive dissonance 11
  strategies for coping with 11
Communication
  dangers in xxiv, xxxvi, 64, 65
  IC models of xiv, 5, 123, 129, 154
  new models of 64
  uncertainty in xx, 8, 58, 65-66, 129
  use of blogs in 66
Complexity xxx, 19, 33-36, 39, 47, 89, 122, 154-155
  confounds Kent paradigm 34-35
Complexity theory 19, 33-34, 39
  analysis inadequate 47, 52, 63, 147, 153, 156
  definition 34-35
Concept map 83, 86-87
  use of 84, 87
  value of 87

INDEX (Continued)

INDEX (Continued)

INDEX (Continued)

INDEX (Continued)

INDEX (Continued)

INDEX (Continued)

COLOPHON

The text of this book was set in Minion Pro, an Adobe Originals typeface designed by Robert Slimbach. It was inspired by classical old-style typefaces of the late Renaissance, a period of elegant, beautiful, and highly readable type designs. Minion Pro exhibits the aesthetic and functional qualities that make text type highly readable, yet is also suitable for display settings. Tables are presented in Neue Helvetica. Helvetica was developed in 1957 by Max Miedinger with Eduard Hoffmann at the Haas'sche Schriftgiesserei (Haas type foundry) of Münchenstein, Switzerland. Haas set out to design a new sans-serif typeface that would compete well in the Swiss market, and be suitable for a wide variety of signage. After further development, the typeface's name was changed by Haas' German parent company to Helvetica (derived from *Confoederatio Helvetica*, the Latin name for Switzerland), to make it more marketable internationally. This Neue Helvetica features a more structurally unified set of heights and widths.

This publication was written on an iMac and a Macbook Pro using Microsoft Word. Typesetting was done using Adobe Systems InDesign. Illustrations were prepared in Omni Systems Omni Graffle and Adobe Photoshop.

Type fonts, as well as iMac, Macbook Pro, Acrobat, InDesign, Omni Graffle, and Photoshop are registered trademarks of their respective companies.

Desktop composition by Joseph Gallagher.

The cover was designed by Bridgette White.

Published by the National Intelligence Press, July 2012. Printed at the Defense Intelligence Agency, Washington, DC.

Notes

Notes

Notes

Notes

Notes

Notes

Notes

Notes